Rayleigh and Lamb Waves

ULTRASONIC TECHNOLOGY

A Series of Monographs

General Editor
Lewis Balamuth
Cavitron Corporation, New York, N.Y.

RAYLEIGH AND LAMB WAVES
Physical Theory and Applications
I. A. Viktorov

Rayleigh and Lamb Waves
Physical Theory and Applications

Igor Aleksandrovich Viktorov
Acoustics Institute
Academy of Sciences of the USSR
Moscow, USSR

Translated from Russian

With a Foreword by
Warren P. Mason
Department of Civil Engineering
and Engineering Mechanics
School of Engineering and Applied Science
Columbia University

Springer Science+Business Media, LLC 1967

Igor Aleksandrovich Viktorov was born in 1932. In 1954 he was graduated from Gorki State University, and from 1954 to 1957 he studied at the Acoustics Institute of the Academy of Sciences of the USSR. He received the degree of Candidate of Physicomathematical Sciences in 1958, and is currently a laboratory director at the Acoustics Institute. He is working on various problems in the propagation of ultrasonic waves in solids, including the direct amplification of ultrasound in semiconducting crystals, and is the author of more than thirty scientific papers.

The original Russian text, first published for the Acoustics Institute of the Academy of Sciences of the USSR in 1966 by Nauka Press, has been updated by the author for the present edition.

Игорь Александрович Викторов
Физические основы применения ультразвуковых волн Рэлея и Лэмба в технике

FIZICHESKIE OSNOVY PRIMENENIYA UL'TRAZVUKOVYKH VOLN RELEYA I LEMBA V TEKHNIKE

RAYLEIGH AND LAMB WAVES: PHYSICAL THEORY AND APPLICATIONS

Library of Congress Catalog Card Number 67-10537

ISBN 978-1-4899-5683-5 ISBN 978-1-4899-5681-1 (eBook)
DOI 10.1007/978-1-4899-5681-1

Originally published by Plenum Press in 1967.
Softcover reprint of the hardcover 1st edition 1967

FOREWORD

Investigations of the transmission of sound waves in solid materials have been carried out in increasing volume over the past 30 years. Most of this increased interest has been due to the fact that such sound waves have been of use in such practical applications as the investigation of the presence of flaws in these materials and in the direct transmission of sound waves in such practical applications as ultrasonic delay lines and in electromechanical filters. The direct application of sound wave transmission originated in the 1930s but received a great impetus during World War II because of the application of delay lines in radar systems and the control and selection of radio frequencies in military communication devices. The ultrasonic inspection method originated with the work of Firestone started during the war. It did not reach its fruition, however, until a number of years after its inception. It has become the most used method for flaw detection.

The original work and most of the early applications were carried out with nearly plane waves which traverse the body of the material and indicate the presence of internal flaws. It was realized that surface flaws could be observed by using surface waves of the Rayleigh type while inspection of plates of material could be carried out by means of Lamb waves. The existence of these waves had been predicted mathematically during the last century and they were shown to have important applications in seismology. However, their application in inspection and wave transmission devices had to wait for a number of investigations on transducers for generating and detecting such waves and theoretical and experimental studies of the effects of flaws on wave transmission. As is evidenced by the complete bibliography of this book, the work was due to many people in many nations. As far as the writer knows, however, this book represents the first synthesis of this knowledge in book form. This book is a very complete account of Rayleigh and Lamb wave propagation and the effect of various types of imperfections on the reflection and propagation of such waves. Both mathematical and experimental techniques are employed. The possible methods for generating and detecting Rayleigh and Lamb waves are discussed. An evaluation of the use of these waves for the determination of the depth and nature of surface working is given. The detection of incipient fatigue cracks appears to be one of the most promising uses.

The use of wave transmission devices such as dispersive and nondispersive delay lines is suggested, but so far only primitive results have been obtained. The full-scale development of such devices still remains to be accomplished.

Columbia University
New York
January, 1967

Warren P. Mason

PREFACE

In 1885, the English scientist Lord Rayleigh (J. Strutt) demonstrated theoretically [1] that waves can be propagated over the plane boundary between an elastic half-space and a vacuum or sufficiently rarefied medium (for example, air), where the amplitude of the waves decays rapidly with depth. These waves, which later came to be named Rayleigh or surface waves, comprise the principal type of wave observed in earth tremors. They have therefore been investigated in detail in seismology (see, for example, [2]).

During the last ten years, Rayleigh waves in the ultrasonic range have found considerable application. They can be used to inspect the state of the surface layer of a sample, i.e., in the detection of surface and near-surface defects in metals, glasses, plastics, and other material (ultrasonic surface flaw detection). The influence of the properties of the surface layer of a sample on the velocity and attenuation of Rayleigh waves permits the latter to be used for the assessment of residual stresses in a surface layer of metal, as well as the thermal and mechanical properties of the surface layer of a sample.

Elastic waves of another type that have found a most diversified application with the evolution of ultrasonics are normal modes, or Lamb waves, named after the scientist who first described them in 1917 [3]. Like ultrasonic Rayleigh waves, Lamb waves are used in ultrasonic flaw detection to determine the elastic and thermoelastic characteristics of sheet samples. Moreover, ultrasonic Rayleigh and Lamb waves are also used in delay lines.

The application of ultrasonic Rayleigh and Lamb waves in technology facilitates the solution of a number of conceptually important problems. For example, it has become feasible to render ultrasonic testing universal, i.e., applicable for components of any shape and size. Heretofore, the volume (longitudinal and transverse) waves used for ultrasonic testing by propagation in solids whose dimension along the wave front is larger than the wavelength have been inapplicable for the testing of thin-walled samples, or for the surface layer of a sample (since the reflection from a defect is masked by reflection from the surface). Ultrasonic Rayleigh and Lamb waves remove these limitations at once. The use of Lamb waves in delay lines has led to the design of a new type of device – dispersive delay lines – with a smoothly regulated time delay.

The first chapter of the present monograph describes the physical attributes of ultrasonic Rayleigh waves, the second chapter describes the physical attributes of ultrasonic Lamb waves, the third chapter gives a cursory account of the main trends in the practical application of these waves.

Research on ultrasonic Rayleigh waves has been carried out at the Acoustics Institute of the Academy of Sciences of the USSR on the initiative of Doctor of Engineering Sciences Yu. M. Sukharevskii. The majority of the author's theoretical investigations have been carried out under the direction of Doctor of Physicomathematical Sciences G. D. Malyuzhinets. The author conveys his sincerest gratitude to him and also expresses his warm appreciation to Doctor of Engineering Sciences L. D. Rozenberg for a discussion of the results of the work, as well as to O. M. Zubova and T. M. Kaekina, junior scientific coworkers of the Acoustics Institute, who collaborated with him in research on the topic in question.

Note to the American Edition

The American edition contains one added section (§9, Chapter II), based on the author's latest investigations on the reflection of Lamb waves from the end section of a plate. Aside from this addition and some minor corrections, the American edition is faithful to the Russian text.

The author will be pleased if the book proves useful to the American reader.

CONTENTS

Rayleigh and Lamb Waves

Chapter I

PHYSICAL PROPERTIES OF ULTRASONIC RAYLEIGH WAVES

§ 1. The Concept of Rayleigh Waves: Their Structure and Properties

As already indicated in the preface, Rayleigh waves represent elastic perturbations propagating near the free boundary of a solid and decaying with depth. The general properties of these waves are described in [4].

Let us consider a plane harmonic Rayleigh wave on the boundary between a solid, isotropic, perfectly elastic half-space and vacuum. Let the half-space occupy the region $z > 0$ (see Fig. 1a), the direction of wave propagation coinciding with the x axis. For the region occupied by the half-space, we introduce the scalar potential ϕ and vector potential $\boldsymbol{\psi}$ of the displacements, so that the particle displacement vector $\mathbf{v}$ is written in the form

$$\mathbf{v} = \operatorname{grad}\varphi + \operatorname{rot}\boldsymbol{\psi}. \tag{I.1}$$

Inasmuch as the wave is plane and the motion does not depend on the coordinate y, only the component of the vector potential along the y axis will have a nonzero magnitude; we denote this component simply by ψ. The potentials ϕ and ψ are called the potentials of longitudinal and shear waves, respectively, and satisfy (for harmonic processes) the following wave equations:

$$\begin{aligned} \frac{\partial^2\varphi}{\partial x^2} + \frac{\partial^2\varphi}{\partial z^2} + k_l^2\varphi &= 0; \\ \frac{\partial^2\psi}{\partial x^2} + \frac{\partial^2\psi}{\partial z^2} + k_t^2\psi &= 0. \end{aligned} \tag{I.2}$$

Here, $k_l = \omega\sqrt{\rho/(\lambda+2\mu)}$ and $k_t = \omega\sqrt{\rho/\mu}$ are the wave numbers for longitudinal and transverse modes, respectively, ω is the circular frequency, λ and μ are the elastic Lamè constants, ρ is the density of the medium. The components U and W of the particle displacement along the x and z axes, respectively, and the stress components σ_{xx}, σ_{zz}, σ_{xz} may be represented in terms of ϕ

and ψ according to the equations

$$U = \frac{\partial\varphi}{\partial x} - \frac{\partial\psi}{\partial z},$$
$$W = \frac{\partial\varphi}{\partial z} + \frac{\partial\psi}{\partial x}; \quad \text{(I.3)}$$

$$\sigma_{xx} = \lambda\left(\frac{\partial^2\varphi}{\partial x^2} + \frac{\partial^2\varphi}{\partial z^2}\right) + 2\mu\left(\frac{\partial^2\varphi}{\partial x^2} - \frac{\partial^2\psi}{\partial x\,\partial z}\right),$$
$$\sigma_{zz} = \lambda\left(\frac{\partial^2\varphi}{\partial x^2} + \frac{\partial^2\varphi}{\partial z^2}\right) + 2\mu\left(\frac{\partial^2\varphi}{\partial z^2} + \frac{\partial^2\psi}{\partial x\,\partial z}\right), \quad \text{(I.4)}$$
$$\sigma_{xz} = \mu\left(2\frac{\partial^2\varphi}{\partial x\,\partial z} + \frac{\partial^2\psi}{\partial x^2} - \frac{\partial^2\psi}{\partial z^2}\right).$$

We seek the solutions to Eqs. (I.2) corresponding to a plane harmonic wave propagating in the positive x direction. For this we let

$$\varphi = F(z)\,e^{i(kx-\omega t)};$$
$$\psi = G(z)\,e^{i(kx-\omega t)}.$$

Substituting these expressions into Eqs. (I.2), we obtain two differential equations for the functions F(z) and G(z):

$$\frac{d^2F(z)}{dz^2} - (k^2 - k_l^2)\,F(z) = 0;$$
$$\frac{d^2G(z)}{dz^2} - (k^2 - k_t^2)\,G(z) = 0.$$

The two linearly independent solutions of each of the above equations are the $\exp(\pm\sqrt{k^2-k_l^2}\,z)$ and $\exp(\pm\sqrt{k^2-k_t^2}\,z)$. We assume *a priori* (as we will see below, this assumption is confirmed) that $k^2 > k_t^2 > k_l^2$. Then the solutions with positive radicals in the exponent will correspond to motion increasing with depth, the solution with negative radicals will correspond to exponentially decaying motion, i.e., a surface wave. Consequently, the expressions for ϕ and ψ assume the form

$$\varphi = Ae^{-qz}e^{i(kx-\omega t)};$$
$$\psi = Be^{-sz}e^{i(kx-\omega t)},$$

where $q^2 = k^2 - k_l^2$; $s^2 = k^2 - k_t^2$; A and B are arbitrary constants.

The conditions of the problem demand also that the stresses σ_{zz} and σ_{xz} go to zero at the boundary of the half-space (plane z = 0). Substituting the expressions for ϕ and ψ into these conditions, we obtain relations linking A, B, and the wave number k. As a result, we finally obtain for ϕ and ψ

$$\begin{aligned} \varphi &= -Ae^{i(kx-\omega t)-qz}; \\ \psi &= iA\frac{2kq}{k^2+s^2}e^{i(kx-\omega t)-sz}. \end{aligned} \tag{I.5}$$

The equation by which k is determined (characteristic equation) has the form

$$4k^2qs-(k^2+s^2)^2=0. \tag{I.6}$$

After a simple transformation, it reduces to the form

$$\eta^6-8\eta^4+8(3-2\xi^2)\eta^2-16(1-\xi^2)=0, \tag{I.7}$$

incorporating the notation $\eta = k_t/k = c/c_t$; $\xi = k_l/k_t = c_t/c_l$ (c_l, c_t are the phase velocities of longitudinal and transverse waves). This relation is sometimes called the Rayleigh equation. It has six roots, the values of which depend only on the Poisson ratio ν for the given elastic medium. The Rayleigh wave corresponds to the root η_R, which lies between zero and one (we will use the subscript R henceforth to indicate quantities pertaining to the Rayleigh wave). It can be shown that for any values of ν corresponding to real media ($0<\nu<0.5$) Eq. (I.7) has one and only one such root. This bears out the validity of our *a priori* assumption and, in addition, demonstrates the possible existence of a Rayleigh wave on the free surface of an elastic half-space. The following is an approximate expression for this root [5]:

$$\eta_R = \frac{0.87+1.12\nu}{1+\nu}. \tag{I.8}$$

As ν varies from 0 to 0.5, the Rayleigh wave phase velocity varies monotonically from $0.87c_t$ to $0.96c_t$. It is readily seen that the Rayleigh wave does not exhibit phase velocity dispersion, since η_R and c_t do not depend on the frequency.

The expressions (I.5) describing a Rayleigh wave show that it consists of two inhomogeneous waves, one longitudinal and one transverse, which propagate along the boundary of the half-space with identical velocities and attenuate with depth according to the laws $\exp(-\sqrt{k_R-k_l^2}z)$ (longitudinal mode)

a

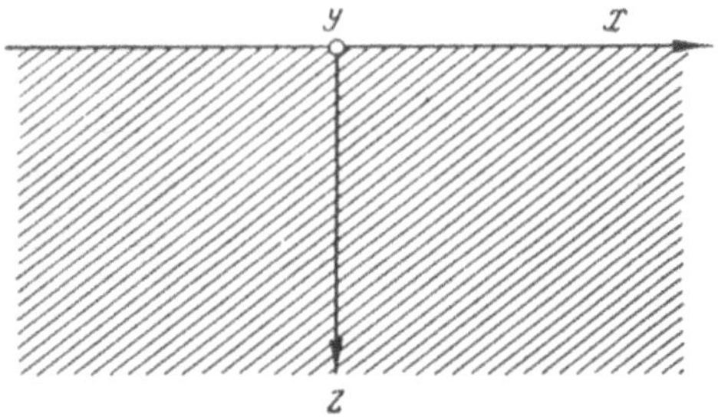

b

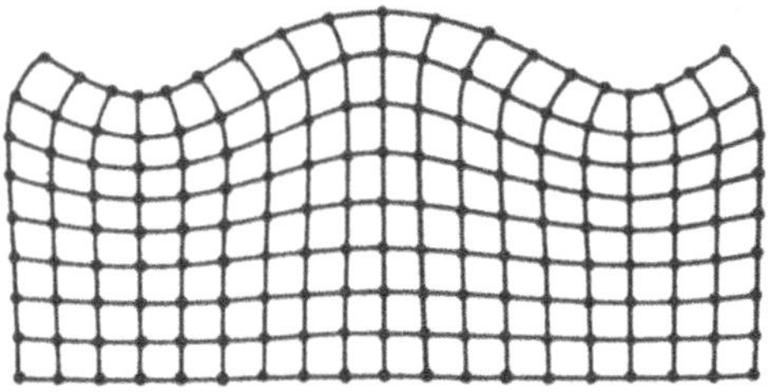

Fig. 1

and $\exp(-\sqrt{k_R^2-k_t^2}z)$ (transverse mode). On the boundary ($z = 0$) these waves cancel the stresses which they produce.

Computing the components of the displacements along the x and z axes by means of the relations (I.3), we obtain (after taking the real parts):

$$U_R = Ak_R\left(e^{-q_R z} - \frac{2q_R s_R}{k_R^2 + s_R^2}\, e^{-s_R z}\right)\sin(k_R x - \omega t);$$
$$W_R = Aq_R\left(e^{-q_R z} - \frac{2k_R^2}{k_R^2 + s_R^2}\, e^{-s_R z}\right)\cos(k_R x - \omega t). \qquad \text{(I.9)}$$

The components of the stresses in the Rayleigh wave may be computed by using the relations (I.4) and (I.5).

The pattern of the displacements in a Rayleigh wave is illustrated in Fig. 1b. The dots indicate the material particles, which in the absence of the wave are situated at equal distances (vertically and horizontally) from one another. Figures 2 and 3 show the dependences, as we have calculated them, of the displacement amplitudes $\hat{U}_R$, $\hat{W}_R$ and stress amplitudes $\hat{\sigma}_{zz}$, $\hat{\sigma}_{xx}$, $\hat{\sigma}_{xz}$ in a Rayleigh wave on the depth. The curves are given in dimensionless form;

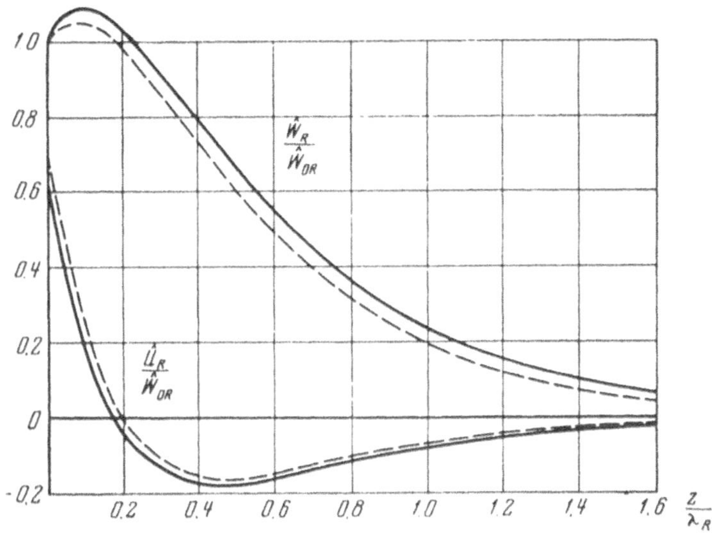
1.0
0.8
0.6
0.4
0.2
0
-0.2
0.2
0.4
0.6
0.8
1.0
1.2
1.4
1.6

Fig. 2

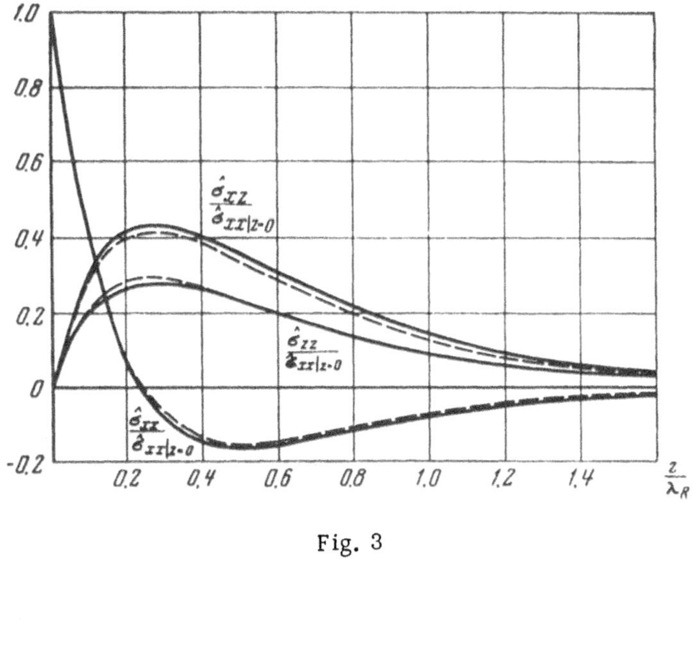
1.0
0.8
0.6
0.4
0.2
0
-0.2
0.2
0.4
0.6
0.8
1.0
1.2
1.4

Fig. 3

the displacement amplitudes are referred to the normal displacement amplitude on the surface $\hat{W}_{0R}$, the stress amplitudes are referred to the amplitude $\hat{\sigma}_x|_{x=0}$ on the surface. The depth is plotted in fractions of the wavelength. The curves were calculated for two values of the Poisson ratio: $\nu = 0.25$ (dashed curves) and $\nu = 0.34$ (solid curves); included between these values are the values of the Poisson ratio for the majority of metals. It is apparent from the graphs that the displacement normal to the surface increases at first, then falls off monotonically with depth, while the displacement parallel to the surface changes sign at a depth of approximately $0.2\lambda_R$. It is also apparent from the graphs that σ_{xx} changes sign, whereas σ_{zz} and σ_{xz} reach a maximum at approximately $z/\lambda_R = 0.3$, then fall off exponentially with depth. The set of curves shown in Figs. 2 and 3 illustrates the localization of the Rayleigh wave in a thin surface layer of thickness $\lambda_R - 2\lambda_R$.

Since the displacement components U_R and W_R in the Rayleigh wave along the x and z axis are shifted in phase by $\pi/2$, the trajectories of the particle motions in the wave are ellipses. For wave propagation in the positive x direction, given our assignment of the coordinate system, the ellipsoidal rotation of the particle at the surface proceeds clockwise; at a depth $z > 0.2\lambda_R$ (when the displacement U_R changes sign) the direction of rotation reverses. The semimajor axes of the ellipses are perpendicular to the boundary of the half-space, the semiminor axes are parallel to the direction of wave propagation. The eccentricity of the ellipses depends on the distance to the surface and on the Poisson ratio of the elastic medium. Table 1, which is borrowed from [6], shows the relative lengths of the semiaxes of the ellipses at different depths for four values of the Poisson ratio ν. The numerator represents the

TABLE 1

$\frac{z}{\lambda_R}$	$\nu=0$	$\nu=0.250$	$\nu=0.333$	$\nu=0.500$
0	$\frac{1.000}{0.772}$	$\frac{1.000}{0.676}$	$\frac{1.000}{0.626}$	$\frac{1.000}{0.540}$
0.250	$\frac{0.745}{-0.007}$	$\frac{0.910}{-0.076}$	$\frac{0.968}{-0.101}$	$\frac{1.120}{-0.158}$
0.500	$\frac{0.432}{-0.104}$	$\frac{0.587}{-0.157}$	$\frac{0.643}{-0.177}$	$\frac{0.812}{-0.206}$
1.000	$\frac{0.109}{-0.044}$	$\frac{0.192}{-0.071}$	$\frac{0.219}{-0.077}$	$\frac{0.339}{-0.099}$

length of the semimajor axis (amplitude $\hat{W}_R$), the denominator the length of the semiminor axis (amplitude $\hat{U}_R$).

In the ultimately simple example of plane harmonic Rayleigh waves in a perfectly elastic isotropic and homogeneous half-space, we have reviewed the most general properties of these waves (velocity, character of the motion in the wave, etc.). In inhomogeneous and anisotropic media, the structure and properties of Rayleigh waves are very complicated and there exist anisotropic media (for example, crystals of the triclinic system) in which Rayleigh waves cannot exist in general. Sometimes Rayleigh waves are interpreted to mean waves not only on the free boundary of a solid, but also surface waves of a more general type, arising at the boundary between a solid and liquid and at the boundary between a system of solid or liquid layers and a solid half-space. At the boundary between solid and liquid half-spaces Rayleigh waves always exist; in all other cases they exist only for certain definite relations between the elastic and geometric parameters of the layers and the solid half-space.

We will investigate within the scope of the present book only ultrasonic Rayleigh waves on the free boundary of a solid or the boundary of a solid with a low-density fluid. Surface Rayleigh waves of a more complex type do not have practical utility at ultrasonic frequencies. We will consider for simplicity plane harmonic ultrasonic Rayleigh waves. The term "ultrasonic" will be dropped in the interest of brevity, especially since most of the properties of ultrasonic Rayleigh waves are inherent in Rayleigh waves of any frequencies.

§ 2. Methods for the Excitation and Reception of Rayleigh Waves

Several methods are available for the excitation of ultrasonic Rayleigh waves. For instance, it is asserted in [7] that an X-cut quartz plate resting on the edge of a rectangular elastic wedge (Fig. 4a) will excite weak surface waves on the faces of the wedge. The optimum excitation is obtained when the plate is inclined at an angle of 45° with respect to the faces.

A method is described in [8] for the excitation of Rayleigh waves by a Y-cut quartz plate which is acoustically bonded to the surface of a solid through a thin film of oil (Fig. 4b). This type of plate stimulates two Rayleigh waves, traveling in the negative and positive x directions with equal amplitudes. As the authors state, the amplitudes are a maximum when the ratio of the plate width 2a to the thickness d is equal to 7 : 1. In this mode of excitation the greater part of the mechanical energy is used up in the radiation of transverse ultrasonic waves into the depth of the solid.

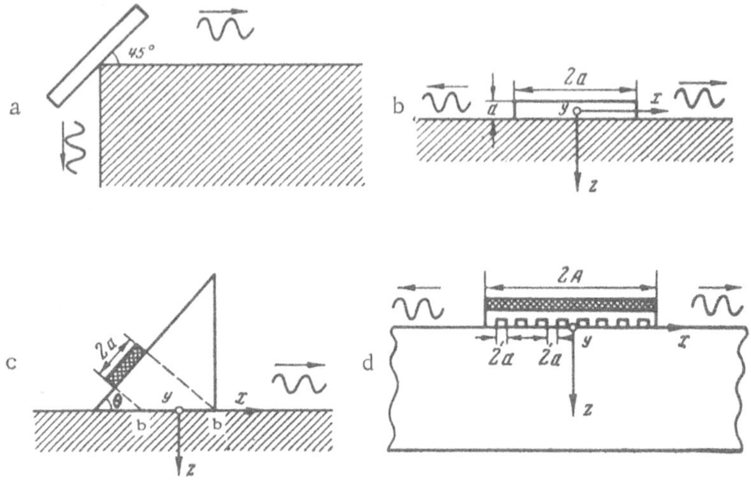

Fig. 4

In [9], a method based on the magnetoacoustic effect was used for the reception of Rayleigh waves from the surface of metal samples; according to this method, when the surface of the sample is set into vibration (under the influence of a Rayleigh wave), eddy currents are created in the sample in a constant magnetic field. The eddy currents produce a magnetic field which varies with the ultrasonic frequency. This method may also be used in principle for the excitation of Rayleigh waves, just as all the excitation methods described herein (with all of their singular characteristics) may be reversed to function in the reception mode.

The most widely employed method of Rayleigh-wave excitation is the so-called wedge or prismatic coupling block method [6, 7], which is based on the conversion of longitudinal waves into Rayleigh modes. In this method (Fig. 4c), a plastic prism or wedge is acoustically coupled through one of its faces to the surface of a solid. A piezoelectric plate is set on the sloping surface of the wedge, where it emits a plane longitudinal wave which impinges at an angle θ to the boundary of the solid and wedge. The angle θ is chosen on the basis of the optimum excitation condition $\sin \theta_R = c_W/c_R$, where c_W is the longitudinal wave velocity in the material of the wedge and c_R is the Rayleigh wave velocity in the sample (it is clear that the material of the wedge must be such that $c_W < c_R$; this is why it is usually made of plastic). This creates at the boundary of the solid and wedge a periodic perturbation with a spatial period equal to the Rayleigh wavelength in the sample. Inasmuch as $\theta_R = \sin^{-1} c_W/c_R$ is greater than the angle of total internal reflection both for longitudinal and for transverse modes, the waves transmitted into the sample are inhomogeneous and the perturbation decays with depth. This perturbation excites a Rayleigh wave which propagates over the surface of the sample in the direction of the positive x axis. The optimum excitation is obtained when the projection of the edge of the wedge onto the sloping face coincides with the leading edge of the piezoelectric plate, as shown in Fig. 4c. In the wedge method, essentially just one Rayleigh wave is excited, the level of the excited volume waves lying 20 to 30 dB lower.

In [10], a somewhat modified wedge method is proposed, with the transformation of a transverse into a Rayleigh wave. Owing to the smaller velocity (by comparison with a longitudinal wave) of the transverse wave, the wedge in this case may be made of metal – for example, brass – rather than of plastic. This has several advantages: simpler fabrication, less wear, better capability for matching of the wedge and sample materials (the latter feature may enhance the efficiency of Rayleigh wave emission and reception severalfold by comparison with the plastic wedge). Some applications of this method are described in [11].

In 1958, A. G. Sokolinskii [12] proposed a method for the excitation of Rayleigh waves by a comblike structure which generates a periodic set of normal-mode perturbations on the surface of a solid (such as a wedge), with a spatial period λ_R (the excitation is less efficient at any spatial period other than λ_R). Such a structure is most simply made in the form of a metal plate with a corrugated, comb-shaped profile consisting of alternating projections and slots of width $\lambda_R/2$ (Fig. 4d) and an X-cut quartz plate resting on top. A comb structure is capable of exciting Rayleigh waves very efficiently in a sample of any material, the only requirement being that its spatial period be equal to λ_R. This gives it an enormous advantage over the wedge method. The drawback of the comb-structure method is the high level of spurious signals associated with the radiation of volume waves by the structure.

We [13] have performed a detailed theoretical and experimental investigation of the four methods of exciting Rayleigh waves: the wedge (prism), comb structure, Y-cut, and X-cut quartz plate. We introduced the last method by analogy with the Y-cut quartz plate. Herein we will describe the main part of this work.

We will consider the solid on whose surface Rayleigh waves are generated to be a homogenous, isotropic, perfectly elastic half-space with a plane free boundary. The dimensions of the emitters along the y axis (Fig. 5) will be assumed infinite and we will consider that the action of the Rayleigh wave emitter on the surface of the solid is equivalent to the action of stresses applied to the free surface of the solid in that portion where the emitter is

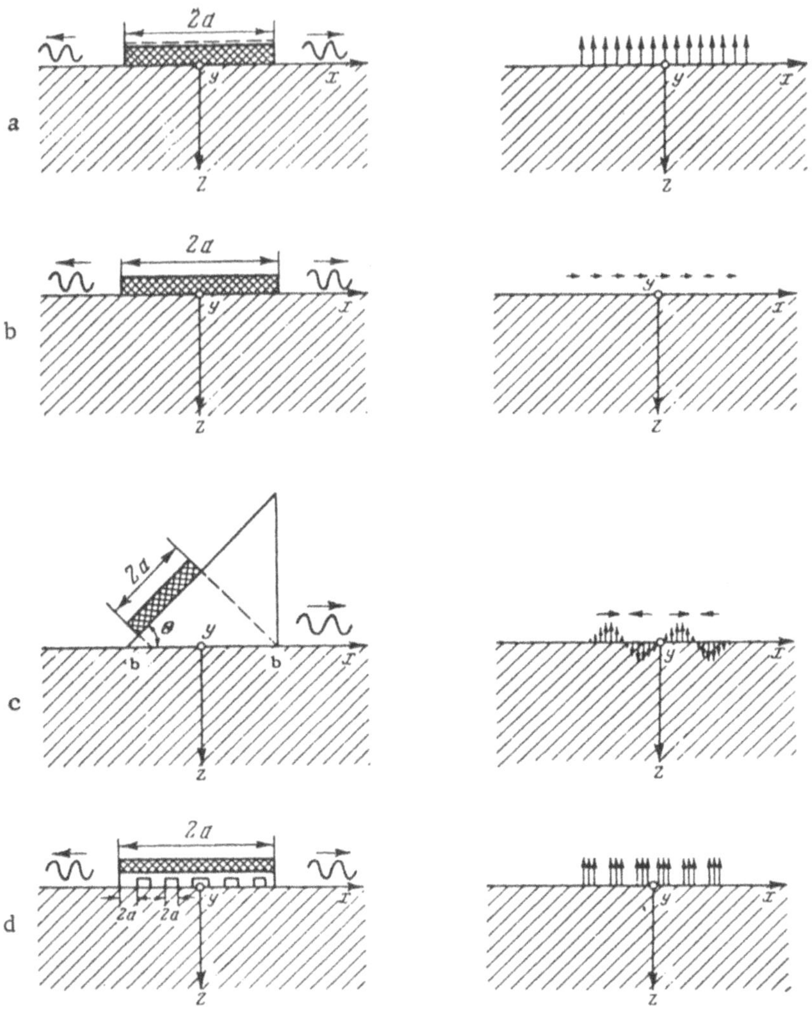

Fig. 5

located. For excitation by X-cut (Fig. 5a) and Y-cut (Fig. 5b) quartz plates, we have respectively, normal and tangential stresses of unit amplitude, distributed uniformly in the region of the surface $|x| \leq a$; for the comb structure (Fig. 5d) we have a periodic set of unit normal stresses; in the wedge method (Fig. 5c) we have a system of normal and tangential stresses applied to the free surface of the solid in the region $|x| \leq a/\cos\theta = b$, which is defined by the geometric boundaries of the beam of longitudinal waves propagated into the wedge. We will assume here that the stresses are equal to those arising with the incidence of a plane longitudinal wave at an angle θ to the boundary of two half-spaces, one of which consists of the wedge material, the second of the material of the solid sample (the longitudinal wave is incident in the first half-space, and its amplitude is assumed to be such that the normal stresses set up in the area perpendicular to its direction of propagation are equal to unity).

Strictly speaking, the "replacement" of Rayleigh wave emitters by stresses is permissible only under the condition of small wave impedances on the part of the emitter materials (i.e., the quartz, material of the comb structure, and material of the wedge) by comparison with the material of the solid, this condition being only approximately met in the majority of practical cases; however, the other limiting case (small wave impedance on the part of the solid), when the Rayleigh wave emitters are replaced by displacements specified on the surface of the solid, is further from realization in practice. We note also that in the wedge method we will neglect the divergence of the longitudinal wave beam in the wedge and the displacement of the reflected beam at the wedge–solid interface.

The time dependence of the stresses will at first be assumed to be harmonic, then later the results obtained will be generalized to the case of pulsed operation.

Given the assumptions described, the investigation of the four methods of Rayleigh wave excitation reduces to an investigation of the vibrations of a half-space with the following stresses on its boundary.

In the case of excitation by an X-cut quartz plate:

$$\begin{aligned} \sigma_{xz|z=0} &= 0 \quad \text{for} \quad |x| < \infty, \\ \sigma_{zz|z=0} &= \begin{cases} \sigma(t) = e^{-i\omega t} & \text{for} \quad |x| \leqslant a, \\ 0 & \text{for} \quad |x| > a. \end{cases} \end{aligned} \tag{I.10}$$

In the case of excitation by a Y-cut quartz plate:

$$\begin{aligned} \sigma_{zz|z=0} &= 0 \quad \text{for} \quad |x| < \infty, \\ \sigma_{xz|z=0} &= \begin{cases} \sigma(t) = e^{-i\omega t} & \text{for} \quad |x| \leqslant a, \\ 0 & \text{for} \quad |x| > a, \end{cases} \end{aligned} \tag{I.11}$$

In the case of excitation by a comb structure:

$$\sigma_{xz|z=0} = 0 \quad \text{for} \quad |x| < \infty,$$

$$\sigma_{zz|z=0} = \begin{cases} \sigma(t) = e^{-i\omega t} & \text{for} \quad |x| \in G, \\ 0 & \text{for} \quad |x| \notin G, \end{cases} \tag{I.12}$$

where $G = \sum_{i=0}^{\pm n} g_i$ is a region consisting of 2n + 1 elementary regions g_i, the boundaries of which are delineated in the plane z = 0 by the straight lines $x_{beg} = (4i+1)a$, $x_{end} = (4i-1)a$.

In the case of wedge excitation:

$$\sigma_{zz|z=0} = \begin{cases} \sigma_z e^{ik_0x}\sigma(t) & \text{for} \quad |x| \leqslant b, \\ 0 & \text{for} \quad |x| > b; \end{cases}$$

$$\sigma_{xz|z=0} = \begin{cases} \sigma_x e^{ik_0x}\sigma(t) & \text{for} \quad |x| \leqslant b; \\ 0 & \text{for} \quad |x| > b, \end{cases} \tag{I.13}$$

where $k_0 = k_W \sin\theta$, $\sigma(t) = e^{-i\omega t}$, σ_x and σ_z are the dimensionless complex stress amplitudes.

The analytical procedure is the same for all modes of excitation. We will therefore limit our presentation to a brief outline of the calculations pertaining to the X-cut quartz-plate excitation method, giving only the final results for the other methods. For the region occupied by the elastic half-space we introduce the potentials ϕ and ψ for longitudinal and shear waves, satisfying the wave equations (I.2). We seek ϕ and ψ in the form of Fourier integrals:

$$\varphi = \int_{-\infty}^{\infty} \varphi(k) e^{i\left(kx + \sqrt{k_l^2 - k^2}z - \omega t\right)} dk;$$

$$\psi = \int_{-\infty}^{\infty} \psi(k) e^{i\left(kx + \sqrt{k_t^2 - k^2}z - \omega t\right)} dk. \tag{I.14}$$

We determine the functions $\phi(k)$ and $\psi(k)$ from the boundary conditions (I.10), writing the expressions for the stresses also in Fourier integral form and expressing the stresses σ_{zz} and σ_{xz} in terms of ϕ and ψ. With regard to the potentials $\phi(k)$ and $\psi(k)$ thus found, from (I.14) we can calculate the displacements and stresses at any point of the half-space, i.e., we determine the total field. Since we are concerned chiefly with Rayleigh modes, i.e., the part of the field which is localized at the surface, we will calculate the surface displacements.

An analysis of the surface displacements also gives a notion as to the set of volume waves excited in each case along with the Rayleigh waves. For the normal and tangential surface displacements we have

$$W_0 = -\frac{ik_t^2}{\pi\mu}\int_{-\infty}^{\infty}\frac{\sqrt{k_l^2-k^2}\sin ka\, e^{ikx}\,dk}{kF(k)};$$
$$U_0 = \frac{i}{\pi\mu}\int_{-\infty}^{\infty}\frac{\left(2k^2-k_t^2+2\sqrt{k_l^2-k^2}\sqrt{k_t^2-k^2}\right)\sin ka e^{ikx}dk}{F(k)}; \qquad \text{(I.15)}$$

where

$$F(k) = 4k^2\sqrt{k_l^2-k^2}\sqrt{k_t^2-k^2} + (k_t^2-2k^2)^2.$$

For the computation of the integrals (I.15), we investigate these integrals in the complex plane of k. In this plane, the integrand functions have the following features: branch points $k = \pm k_{l,t}$ of the radicals $\sqrt{k_{l,t}^2-k^2}$ and the simple poles $k = \pm k_R$, corresponding to the simple roots of the function F(k) (regarding the latter, see, e.g., [14)]. In order to render the functions $\sqrt{k_{l,t}^2-k^2}$ single-valued, we form a Riemann two-sheeted surface from the two sheets of the k plane, making the branch cuts as shown in Fig. 6. We refer to the upper sheet of the Riemann surface as that one for which the signs of the radical $\sqrt{k_{l,t}^2-k^2}$ correspond to satisfaction of the extinction principle [15] for the solution (I.14). The paths of integration in (I.14) and (I.15) are taken along the real axis of the upper sheet. We compute the surface displacements for the region outside the plate ($|x| > a$). Taking the paths of integration of (I.15) from the real axis of the upper sheet to positive or negative (depending on the sign of x) imaginary infinity, we reduce the integrals (I.15) to residues at the points $k = \pm k_R$ and integrals along the borders of the cuts (see Fig. 6) (the direction in which the poles $k = \pm k_R$ lying on the real axis are bypassed is determined by the insertion of an infinitesimal attenuation into the half-space, so that the pole is just displaced from the real axis – see the arrows in Fig. 6). For large x [$k_{l,t}(x-a) \gg 1$], the integrals along the borders of the cuts may be computed approximately by the method of steepest descent, whereupon for W_0 and U_0 we have the expressions (I.16) below. The corresponding equations for all the modes of excitation are presented herewith.

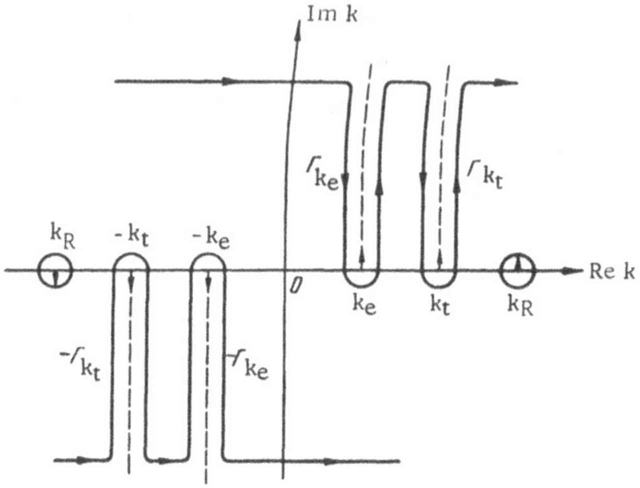

Fig. 6

In the case of excitation by an X-cut quartz plate $[k_{l,t}(x-a) \gg 1]$:

$$\begin{aligned} W_0 &= -iA\frac{q_R}{k_R}\frac{\sin k_R a}{k_R}e^{ik_R x} + B_1\frac{\sin k_l a}{k_l}\frac{e^{ik_l x}}{k_l^{3/2}x^{3/2}} + \\ &+ C_1\frac{\sin k_t a}{k_t}\cdot\frac{e^{ik_t x}}{k_t^{3/2}x^{3/2}} + 0\left[\frac{1}{k_{l,t}^2 x^2}\right]; \\ U_0 &= A\frac{k_R^2+s_R^2}{2k_R^2}\frac{\sin k_R a}{k_R}e^{ik_R x} + B_2\frac{\sin k_l a}{k_l}\frac{e^{ik_l x}}{k_l^{3/2}x^{3/2}} + \\ &+ C_2\frac{\sin k_t a}{k_t}\frac{e^{ik_t x}}{k_t^{3/2}x^{3/2}} + 0\left[\frac{1}{k_{l,t}^2 x^2}\right]. \end{aligned} \tag{I.16}$$

In the case of excitation by a Y-cut quartz plate $[k_{l,t}(x-a) \gg 1]$:

$$\begin{aligned} W_0 &= -A\frac{k_R^2+s_R^2}{2k_R^2}\frac{\sin k_R a}{k_R}e^{ik_R x} + B_3\frac{\sin k_l a}{k_l}\frac{e^{ik_l x}}{k_l^{3/2}x^{3/2}} + \\ &+ C_3\frac{\sin k_t a}{k_t}\frac{e^{ik_t x}}{k_t^{3/2}x^{3/2}} + 0\left[\frac{1}{k_{l,t}^2 x^2}\right]; \end{aligned}$$

$$U_0 = -iA\frac{s_R}{k_R}\frac{\sin k_R a}{k_R}e^{ik_R x} + B_4\frac{\sin k_l a}{k_l}\frac{e^{ik_l x}}{k_l^{3/2}x^{3/2}} + C_4\frac{\sin k_t a}{k_t}\frac{e^{ik_t x}}{k_t^{3/2}x^{3/2}} + 0\left[\frac{1}{k_{l,t}^2 x^2}\right]. \tag{1.17}$$

In the case of excitation by a comb structure $[k_{l,t}x-(4n+1)k_{l,t}a>>1]$:

$$W_0 = -iA\frac{q_R}{k_R}\sum k_R a_n e^{ik_R x} + B_5\sum k_l a_n\frac{e^{ik_l x}}{k_l^{3/2}x^{3/2}} + C_5\sum k_t a_n\frac{e^{ik_t x}}{k_t^{3/2}x^{3/2}} + 0\left[\frac{1}{k_{l,t}^2 x^2}\right];$$

$$U_0 = A\frac{k_R^2+s_R^2}{2k_R^2}\sum k_R a_n e^{ik_R x} + B_6\sum k_l a_n\frac{e^{ik_l x}}{k_l^{3/2}x^{3/2}} + C_6\sum k_t a_n\frac{e^{ik_t x}}{k_t^{3/2}x^{3/2}} + 0\left[\frac{1}{k_{l,t}^2 x^2}\right]. \tag{I.18}$$

In the case of excitation by a wedge $[k_{l,t}(b-x)>>1]$:

$$W_0 = -A\left(i\frac{q_R}{k_R}\sigma_z + \frac{k_R^2+s_R^2}{2k_R^2}\sigma_x\right)\frac{\sin(k_0-k_R)b}{k_0-k_R}e^{ik_R x} + (B_7\sigma_z + B_8\sigma_x)\frac{\sin(k_0-k_l)b}{k_0-k_l}\cdot\frac{e^{ik_l x}}{k_l^{3/2}x^{3/2}} + (C_7\sigma_z + C_8\sigma_x)\frac{\sin(k_0-k_t)b}{k_0-k_t}\frac{e^{ik_t x}}{k_t^{3/2}x^{3/2}} + 0\left[\frac{1}{k_{l,t}^2 x^2}\right];$$

$$U_0 = A\left(\frac{k_R^2+s_R^2}{2k_R^2}\sigma_z - i\frac{s_R}{k_R}\sigma_x\right)\frac{\sin(k_0-k_R)b}{k_0-k_R}e^{ik_R x} + (B_9\sigma_z + B_{10}\sigma_x)\frac{\sin(k_0-k_l)b}{k_0-k_l}\cdot\frac{e^{ik_l x}}{k_l^{3/2}x^{3/2}} + (C_9\sigma_z + C_{10}\sigma_x)\frac{\sin(k_0-k_t)b}{k_0-k_t}\frac{e^{ik_t x}}{k_t^{3/2}x^{3/2}} + 0\left[\frac{1}{k_{l,t}^2 x^2}\right]. \tag{I.19}$$

The following notation is introduced here:

$$\sum k_{l,t,R} a_n = \frac{1}{k_{l,t,R}} [\sin(4n+1) k_{l,t,R} a -$$

$$-\sin(4n-1) k_{l,t,R} a + \sin(4n-3) k_{l,t,R} a - \sin(4n-5) k_{l,t,R} a +$$

$$+ \ldots + \sin k_{l,t,R} a];$$

$$A = \frac{k_t^2 q_R s_R}{2\mu\left[k_R^2 (q_R^2 + s_R^2) - 2q_R s_R (k_R^2 + s_R^2 - q_R s_R)\right]};$$

$B_1 \ldots B_{10}, C_1 \ldots C_{10}$ are complex constants depending on the Poisson ratio ν and Lamè constant μ of the elastic medium.

It is clear from the equations presented above that in each method of excitation the expressions for the surface displacements far from the region where the stresses are applied consist of a sum of a series of terms, each of which, as inferred from their phase multipliers, corresponds to a definite wave type. The first terms of Eqs. (I.16)-(I.19) correspond to Rayleigh modes, the second and third to longitudinal and transverse modes propagating to the right from the regions of stress application. The amplitudes of the surface displacements in the volume waves are considerably less [by a factor of $1/(k_{l,t}x)^{3/2}$] than the corresponding Rayleigh wave amplitudes, and decay with distances along the surface of the half-space according to the law $(k_{l,t}x)^{-3/2}$. This rapid damping of the amplitudes is attributable to the fact that the main portion of the energy of the longitudinal and transverse waves is radiated into the depth of the half-space, rather than over its free surface. We point out that for large negative x, the amplitudes W_0 and U_0 differ from the amplitudes in (I.16)-(I.19) only in sign. An exception is the wedge,case, where the radiation in the -x direction is less than in the +x direction. In particular, for $k_0 \simeq k_R$ the amplitude of a Rayleigh wave traveling in the -x direction is insignificantly small ($^1/_2 k_R b$ times the amplitude in the +x direction). Analyzing the expressions for the amplitudes of the surface displacements W_{0R} and U_{0R} in the Rayleigh mode, we arrive at the following conclusions: (1) In excitation by X- and Y-cut quartz plates the Rayleigh wave amplitude depends sinusoidally on the width 2a of the plate [see the first terms in (I.16) and (I.17) and Fig. 7, which shows theoretical (1) and experimental (2) curves]. The elastic half-space is assumed here and below to be "aluminum"; (2) in the excitation of Rayleigh waves by a comb structure or wedge, when there exists on the surface of the half-space a region with periodic perturbations, the Rayleigh wave amplitude depends very strongly on the spatial period of these perturbations, i.e., on the angle θ and dimension 4a, respectively [see Fig. 8,

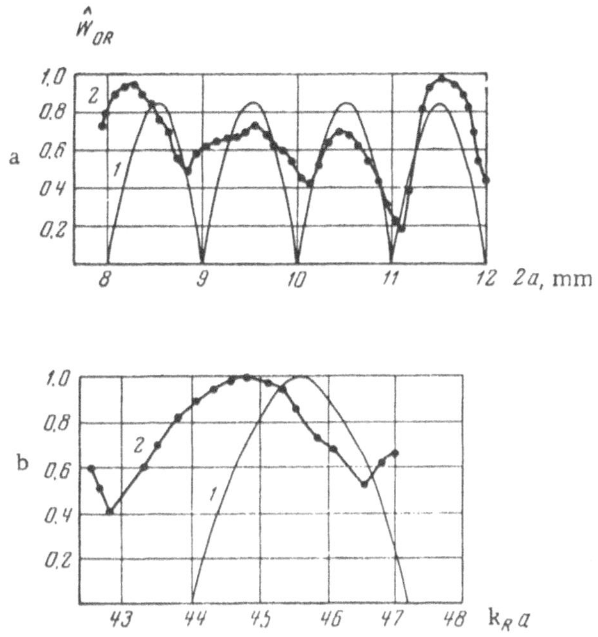

Fig. 7

which shows theoretical (1) and experimental (2) curves: the theoretical curves are calculated according to Eqs. (I.18) and (I.19) for the case of a comb structure in which the number of unit cells is equal to 19(n = 9) and for a polystyrene wedge with quartz plate of width $2a = 7.66\lambda_W$]. Analyzing Eqs. (I.18) and (I.19) we are readily convinced that the sharpness of the maxima in the curves of Fig. 8 is proportional to the size of the perturbation region, i.e., to the number of cells m and the quantity 2b; (3) in the excitation of Rayleigh waves by X- and Y-cut quartz plates, the maximum excitation efficiency, which occurs at $2a = \lambda_R/2$, $3\lambda_R/2$, $5\lambda_R/2$, does not depend on the width of the plate 2a. In excitation by a comb structure and wedge, the maximum Rayleigh wave amplitude, corresponding to the cases $4a = \lambda_R$ and $\theta \simeq \sin^{-1} k_R/k_W$ $(k_0 \simeq k_R)$, as implied by (I.18) and (I.19), is directly proportional to the dimension of the emitters in the direction of Rayleigh wave propagation and theoretically it may be unbounded.

From the results obtained above, it is also possible to make certain conclusions regarding the pulsed mode of excitation. We will consider the most frequently met case, when the stress pulse used to "replace" the emitter has a rectangular shape and a sinusoidal carrier, i.e., the function $\sigma(t)$ [see (I.10)-

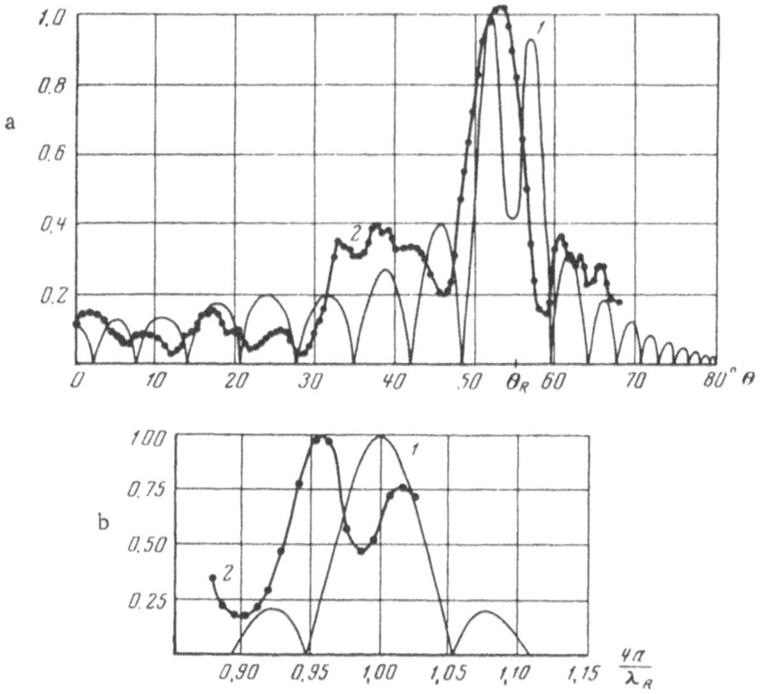

Fig. 8

(I.13)] is equal to

$$\sigma(t) = \begin{cases} e^{-i\omega_0 t} & \text{for } |t| \leqslant \tau \\ 0 & \text{for } |t| > \tau \end{cases}$$

(2τ is the pulse duration, ω_0 the carrier frequency), where at the initial and final instants of time the stresses are equal to zero (Re $e^{\pm i\omega_0 \tau} = \cos \omega_0 \tau = 0$). Representing σ(t) in Fourier integral form, we obtain, after a few simple calculations, a solution for the pulsed mode from a superposition of the solutions (I.16)-(I.19) for the continuous mode. Figure 9 shows oscillograms of the mechanical stress σ_{zz} or σ_{xz} (Fig. 9a) and normal surface displacement in a Rayleigh wave for all modes of excitation: b, c) X- and Y-cut quartz plates; d) comb structure; e) wedge; the curves are calculated for excitation of X- and Y-cut quartz plates on the assumption that $\sin k_R a = 1$, for excitation by a comb structure with m = 3, and for the wedge method with $k_0 = k_R$ ($\theta = \theta_R$). As apparent from Fig. 9, in all methods of excitation except the wedge method, the Rayleigh wave pulse has a large duration (by comparison with the stress pulse) and altered shape [a step function in the case of excitation by X- and Y-cut quartz plates (see Figs. 9b, c) and trapezoidal in the case of excitation by a comb structure (see Fig. 9d). This change in shape and duration is related to the process of vibration buildup in the region of application of the stresses, the process requiring a time $2L/c_R$, where L is the dimension of the emitter in the direction of Rayleigh wave propagation. In the wedge method, for $k_0 = k_R$ the buildup process occurs with a gradual onset of stresses in the region $|x| \leq b$. The corresponding stress pulse created by the longitudinal waves in a wedge occurs at the initial instant on the left face of the region $|x| \leq b$ and then propagates toward the right with a velocity c_R. At the moment the stresses appear over the entire region $|x| \leq b$, the vibrations also stabilize over the entire region. Consequently, the Rayleigh wave pulse repeats the shape of the stress pulse.

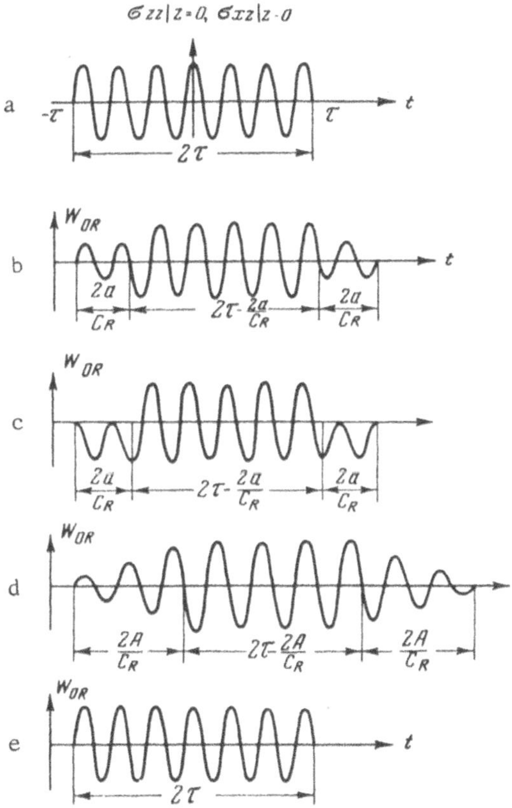

Fig. 9

Experimental testing of the theoretical calculations was carried out on a pulsing apparatus consisting of an electrical rectangular pulse generator with a 2.5-3.5 Mc carrier frequency, a resonance amplifier, and indicator (25-I synchroscope). The acoustic line for the Rayleigh waves consisted of aluminum bars of rectangular cross section and dimensions 25 x 65 x 500 mm. In all of the experiments, the Rayleigh waves were received by the wedge method. The measurements were taken over an electrical-pulse duration of 2 to 20 μsec. The readings of the pulse amplitudes on the indicator were taken from the middle part of the pulse, corresponding to the vibrational steady state, to which Eqs. (I.16)-(I.19) are applicable.

To determine the dependence of the amplitude W_{0R} of the normal surface displacement in the Rayleigh wave on the width of the X-cut quartz plate, we varied the length of the area over which acoustic coupling was established between the X-cut quartz plate and the bar. To record the analogous dependence for a Y-cut quartz plate, we bonded the latter with ceresin to the surface of the bar and varied the electrical pulse carrier frequency to the plate, in effect, expressing the size of the plate in terms of wavelengths λ_R. The resultant curves for $\hat{W}_{0R}(2a)$ and $\hat{W}_{0R}(k_R a)$ are indicated by the dots in Fig. 7. The curves are normalized: the mean height of the maxima on the experimental curve for $\hat{W}_{0R}(2a)$ is set equal to the height of the theoretical maxima, while for the experimental and theoretical curves for $\hat{W}_{0R}(k_R a)$ the heights are assume to be identical. As evident from Fig. 7, the experimental curves disclose qualitatively the same periodic behavior in the dependence of the amplitude on the plate width as the theoretical curves. Quantitatively, however, the curves differ appreciably.

For the investigation of the dependence of the Rayleigh wave amplitude in the wedge method on the spatial period of the perturbations set up by the emitter, i.e., on the angle θ, we used a circular polystyrene sector with an apex angle of 90° and, contacting with the bar and moving along the sector, a sliding contact with an X-cut quartz plate mounted on it. The sliding contact could be fixed in any position on the sector, thus allowing the beam of longitudinal waves to be directed onto the surface of the bar at angles within the limits 0-80°. In recording the analogous dependence $\hat{W}_{0R}(4a/\lambda_R)$ for a comb structure, we used an aluminum plate with a comb-shaped profile and, mounted on it, an X-cut quartz plate. The parameter $4a/\lambda_R$ was varied by changing the carrier frequency of the electrical pulse supplied to the emitter. Acoustic coupling was established between the polystyrene sector and bar by a solid film of BF-4 cement, all other acoustic couplings being made with castor oil. The experimental curves for $\hat{W}_{0R}(\theta)$ (Fig. 8a) and $\hat{W}_{0R}(4a/\lambda_R)$ (Fig. 8b) are "reduced" to the theoretical: the mean height of the first three maxima of the experimental curve for $\hat{W}_{0R}(\theta)$ is taken equal to the mean height of the theoretical, while in the case of the curves for $\hat{W}_{0R}(4a/\lambda_R)$, the heights of the principal maxima are made equal. It is apparent from the figures that the qualitative behavior of the experimental and theoretical curves is the same. A characteristics feature of the experimental curve in Fig. 8a (as compared with the theoretical) is the more distinct principal maximum, the position of which satisfies (within the limits of experimental error) the condition $k_0 = k_R$ $(\theta = \theta_R = \sin^{-1} k_R/k_W)$, which is consistent with the data of [6].

We also investigated the dependence of the Rayleigh wave amplitude in resonance methods of excitation on the dimension of the emitters in the direction of wave propagation during spatial resonance, i.e., for $4a = \lambda_R$ and $\theta = \theta_R$. To record the dependence $\hat{W}_{0R}(m)$, on the comb-shaped plate, we varied the number of tines in contact with the bar. The dependence $\hat{W}_{0R}|_{\theta=\theta_R}$ $(2b/\lambda_R)$ was investigated by varying the width 2a of the surfaces of acoustic coupling between the wedge and the quartz plate bonded to it. These dependences are shown in Fig. 10. Here also are indicated (by thin lines) the corresponding theoretical curves, constructed on the basis of (I.18) and (I.19), taking into account the experimentally measured attenuation of the Rayleigh wave during propagation beneath the emitter, a factor which was not taken into account in (I.18) and (I.19). In constructing the theoretical curves, the amplitude factor was so chosen that the slopes of the experimental and theoretical curves were identical near the origin. As evident from Fig. 10, the theoretical and experimental curves for $\hat{W}_{0R}|_{\theta=\theta_R}(2b/\lambda_R)$ are in satisfactory agreement, while the experimental curve for $\hat{W}_{0R}(m)$ tends to "saturation" theoretically more rapidly.

The experiments described confirm the fact that the amplitude of the Rayleigh wave depends considerably on the dimension of the emitter in the

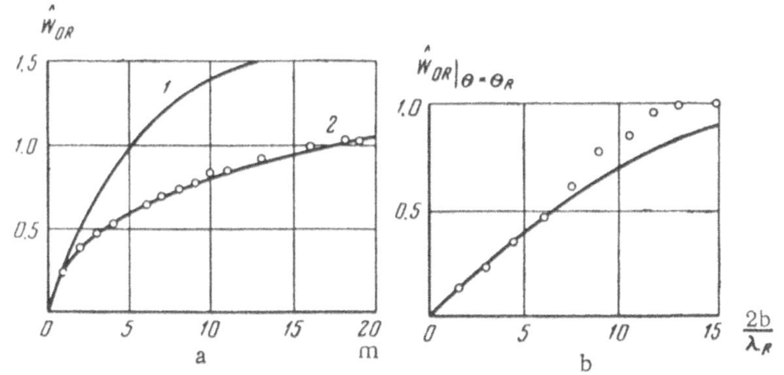

Fig. 10

TABLE 2

Mode of excitation	K, dB
Rayleigh wave excitation by X-cut quartz plate with castor oil coupling layer	56.8
Rayleigh wave excitation by Y-cut quartz plate with ceresin coupling layer	56.4
Rayleigh wave excitation by Y-cut quartz plate with 7:1 ratio of the width 2a to the thickness d ($\sin k_R a \neq 1$); ceresin coupling layer	57.5
Rayleigh wave excitation by an aluminum plate with comb profile ($m = 19$, $2A = 18.5\lambda_R$) with X-cut quartz plate attached to the top; castor oil films as coupling layers	41.1
Rayleigh wave excitation by a polystyrene wedge ($2b = 17\lambda_R$) with BF-4 cement layer	51.4
Excitation and reception of longitudinal waves by X-cut quartz plates with castor oil coupling layers	35.3
Excitation and reception of transverse waves by Y-cut quartz plates with ceresin coupling layers	35.7

direction of wave propagation for all the modes of excitation, and in resonance methods it also depends on the spatial structure of the emitter.

We measured the conversion factors K for the transformation of an electrical pulse into an acoustical pulse and back again for all the modes of excitation.* The source was a polystyrene wedge. The values of K corresponding to the peak (in every case) Rayleigh wave amplitudes (i.e., measured for $\sin k_R a = 1$, $4a = \lambda_R$, $\theta = \theta_R$) are given in Table 2. For comparison, the values of K are also given for volume waves. All of the measurements were made at a frequency of $\simeq 2.7$ Mc.

It is evident from the table that the comb-structure method is the most effective, although even for this method the efficiency is considerably lower than in the excitation and reception of volume waves. We point out that the data of Table 2, Fig. 7b, and Eq. (I.17), in our opinion, refute the assertion of the authors of [8] that the amplitude of a Rayleigh wave emitted by a Y-cut quartz plate is a maximum for a 7:1 ratio of the plate width 2a to the thickness d.

Photographs of Rayleigh wave pulses are shown in Fig. 11 for excitation by X-cut (b) and Y-cut (c) quartz plates, a comb structure (d), and a wedge (e).

*K is interpreted, as usual, to mean the amplitude ratio of the electrical potential on the emitter to the emf developed by the receiver.

The photographs on the left were made with an electrical pulse duration $2\tau = 5$ μsec, those on the right were made with a pulse duration $2\tau = 10$ μsec. The photographs were obtained with the following emitter dimensions in the direction of Rayleigh wave propagation: $2a = 15$, $2A = 18$, $2b = 16$ mm, with maximum Rayleigh wave amplitude in each case, i.e., corresponding to the conditions $\sin k_R a = 1$, $4a = \lambda_R$, $\theta = \theta_R$. For an estimate of the distortion brought into the Rayleigh wave pulse by the quartz plate and amplifier, the Rayleigh wave pulse was not compared with the electrical voltage pulse, but with a pulse of longitudinal waves emitted and received by an X-cut quartz plate (see Fig. 11a, where, as before, on the left $2\tau = 5$ μsec, on the right $2\tau = 10$ μsec).

As apparent from Fig. 11, in correspondence with the theoretical data (see Fig. 10) the Rayleigh wave pulse is elongated and alters its shape (by comparison with the electrical pulse) in all the modes of excitation except the wedge method, wherein both the duration and shape of the pulse are preserved.

As remarked above, most of the experimental data cited agree only qualitatively with their theoretical counterparts. This is clearly due primarily to the lack of correspondence between the theoretical idealizations and the true state of affairs, namely in the following:

1) The effect of the emitter on the surface of a solid is not equivalent to the effect of surface stresses;

2) the distribution of the stresses set up by real Rayleigh wave emitters of the given type differ from the idealized versions (see Fig. 5) due to the nonpiston-like vibrations of rectangularly cut quartz plates (see [5]), irregularities of the intermediate coupling layer, divergence and attenuation of the beam of longitudinal waves in the material of the wedge (for the wedge method), and imperfections in the fabrication of the comb-structured aluminum plate (nonuniform width of the projections and slots, etc.) for the comb-structure method;

3) during propagation over the acoustic coupling surface between the emitter and bar, the Rayleigh waves are attenuated (a fact which is accounted for only in the construction of the theoretical curves in Fig. 10), as a result of which the contribution of dissimular regions of the emitter contact surface to the formation of the Rayleigh wave is nonuniform. Moreover, the discrepancy between the experimental and theoretical curves could be due to experimental errors, specifically the error in measuring the frequency ω_0, which affects the position of the maxima and minima of the experimental curves for $\hat{W}_{0R}(k_R a)$ and $\hat{W}_{0R}(4a/\lambda_R)$ in Figs. 7 and 8.

It is clear from the investigation described above that in the excitation of Rayleigh waves the dependence of their amplitude on the parameters of the

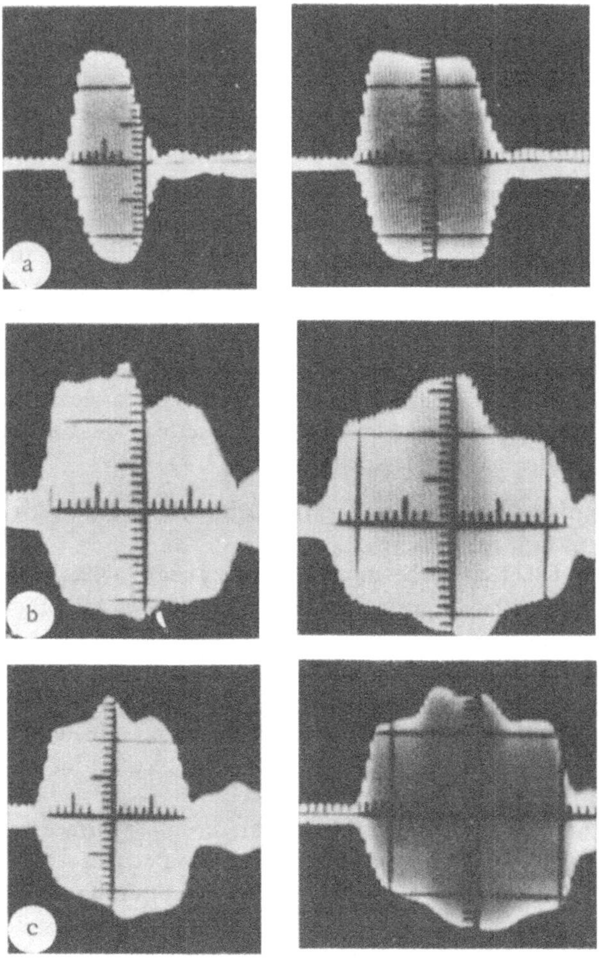

Fig. 11 (a, b, c)

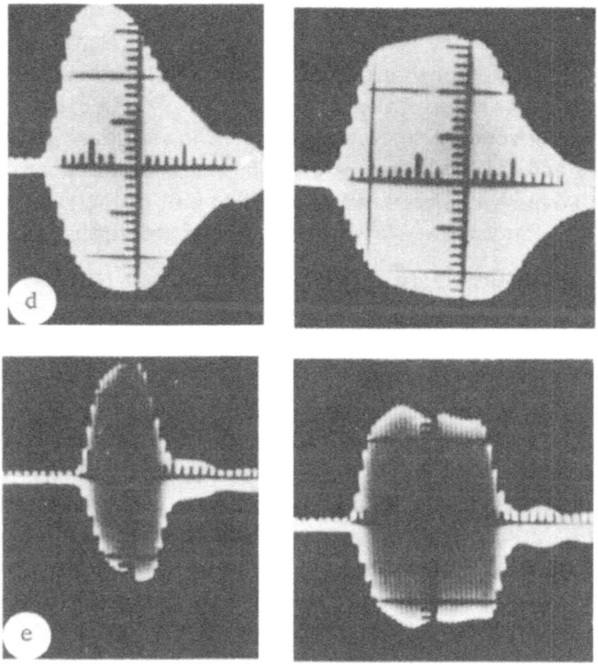

Fig. 11 (d, e)

emitter has a number of characteristic features over and above the case of volume wave emission. These features are described by Eqs. (I.16)-(I.19) and must be given due regard in the selection of a particular excitation mode, as well as in the design of a suitable emitter.

§ 3. Attenuation of Rayleigh Waves

Since Rayleigh waves do not penetrate into the depth of a solid, their amplitude decays with distance as $1/\sqrt{k_R R}$ (where R is the distance from the source) due to divergence of the wave beam emitted by the source. This decay is like that of cylindrical waves, i.e., it occurs more slowly than in the case of volume waves, where the analogous decay law is $1/k_{l,t}R$. This is precisely why Rayleigh waves are the principal type of waves recorded in earthquakes.

The amplitude decay of Rayleigh waves due to absorption and scattering of acoustic energy must be inherent in the features typical of longitudinal and transverse waves, since the Rayleigh wave, as already mentioned, is a combination of these waves.

The absorption and scattering of Rayleigh waves at ultrasonic frequencies have been only rather meagerly investigated. The attenuation of volume (longitudinal and transverse) ultrasonic waves, on the other hand, has been explored in detail (see, e.g., the monograph [5]). Consequently, the few papers that are devoted to the investigation of Rayleigh wave attenuation are concerned by and large with the relation between the attenuation of surface and volume waves. An equation is derived in [16, 17], linking the attenuation factors of such waves. We will give a brief description of its derivation.

Let us consider an elastic medium with losses, where

$$\bar{k}_l = k_l' + ik_l'';$$
$$\bar{k}_t = k_t' + ik_t'';$$
$$\bar{k}_R = k_R' + ik_R''.$$

Let the attenuation of longitudinal and transverse waves be small ($k_l' \gg k_l''$, $k_t' \gg k_t''$) and identical at all points of the medium. As will become apparent shortly, the attenuation of Rayleigh waves in this case is also small and identical at all points of the medium. We write the complex wave numbers for longitudinal, transverse, and Rayleigh modes in the form

$$\begin{aligned} \bar{k}_l &= k_l'(1 + i\alpha); \\ \bar{k}_t &= k_t'(1 + i\beta); \\ \bar{k}_R &= k_R'(1 + i\gamma). \end{aligned} \tag{I.20}$$

Here $\alpha = k_l''/k_l'$; $\beta = k_t''/k_t'$; $\gamma = k_R''/k_R'$ are small real corrections numerically equal to the attenuation factors referred to 2π for longitudinal, transverse, and Rayleigh modes per unit wavelength in the corresponding mode.

To determine $\bar{k}_R$ we turn to the characteristic equation, which, for complex wave numbers has the form

$$\bar{\eta}^6 - 8\bar{\eta}^4 + 8\bar{\eta}^2(3 - 2\bar{\xi}^2) - 16(1 - \bar{\xi}^2) = 0, \tag{I.21}$$

where

$$\bar{\eta}^2 = \bar{k}_t^2/\bar{k}_R^2; \quad \bar{\xi}^2 = \bar{k}_l^2/\bar{k}_t^2.$$

The given equation is obtained from Eq. (I.7) by replacement of the real η and ξ with their corresponding complex values. We substitute the wave numbers (I.20) into the expression for $\bar{\eta}$ and $\bar{\xi}$ and then the latter into Eq. (I.21). In this equation, we retain terms of order one and α, β, γ, rejecting terms of order α^2, β^2, γ^2, and higher. Equating the real and imaginary parts of the truncated equation to zero, we obtain two equations. The first determines the real part of the Rayleigh wave number in the given medium in terms of the real parts of the longitudinal and transverse wave numbers, and from the second we have

$$\gamma = A\alpha + (1 - A)\beta. \tag{I.22}$$

Here

$$A = \frac{16\xi'^2 (1 - \eta'^2)}{\eta'^2 (3\eta'^4 - 16\eta'^2 - 16\xi'^2 + 24)},$$

$$\xi' = \frac{k_l'}{k_t'}, \quad \eta' = \frac{k_t'}{k_R'}.$$

Consequently, the Rayleigh wave attenuation factor per wavelength is a linear combination of the analogous longitudinal and transverse wave attenuation factors. The quantity A depends only on the Poisson ratio ν. Our analytic curves for the dependences of A and 1−A [the coefficient of β in (I.22)] on ν are shown in Fig. 12. It is evident from the graph that for all materials the relative contribution of β in Eq. (I.22) is greater than that of α, i.e., for approximately the same values of α and β the Rayleigh wave attenuation factor γ is governed primarily by the transverse attenuation factor β.

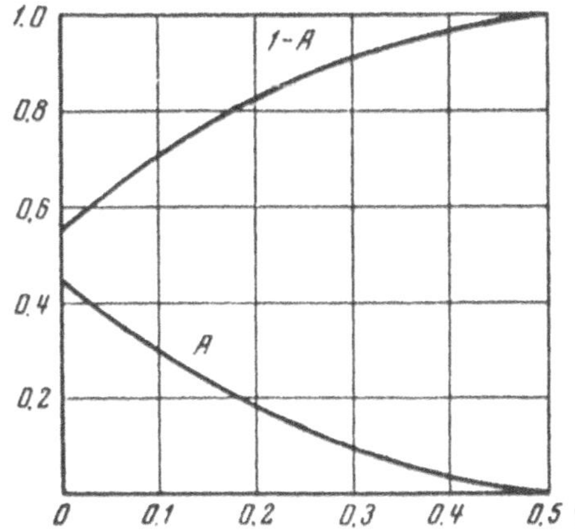

Fig. 12

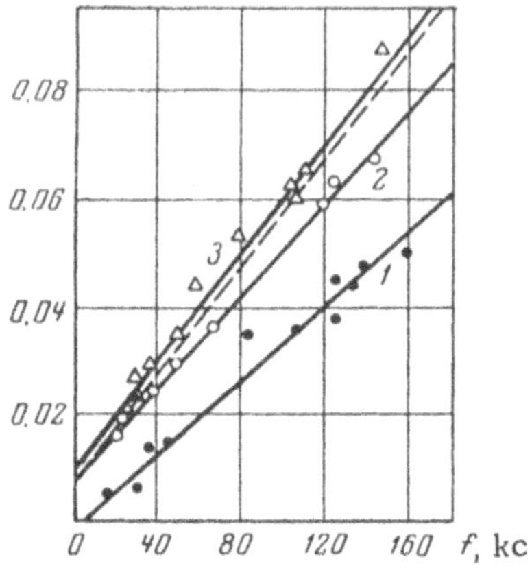

Fig. 13

For example, in the case of steel $\gamma = 0.89\beta + 0.11\alpha$, for aluminum $\gamma = 0.93\beta + 0.07\alpha$, for rubber ($\nu = 0.5$) $\gamma = \beta$.

The results of experiments testing Eq. (I.22), performed on plexiglass samples by the pulsing method in the frequency range 20 to 180 kc, are given in [16]. These experiments provide good confirmation of the relation between the attenuation factors in longitudinal (dots in Fig. 13), transverse (circles), and Rayleigh (triangles) modes, as given by Eq. (I.22) (the solid lines in Fig. 13 are linear approximations obtained for the experimental dependences by the method of least squares; the dashed line represents the analytic dependence of the Rayleigh wave attenuation factor).

The attenuation factors for Rayleigh and longitudinal ultrasonic waves in Dural, MA-3 alloy, and fused quartz at 2.5-8 Mc were measured in [9]. The Rayleigh and longitudinal attenuation factors turned out to be of the same order of magnitude, which in some measure corroborates Eq. (I.22).

In [18], a verification of the relation (I.22) was made in the megacycle range for three solid materials of diverse types: a metal (Dural), a glass (mirror glass), and a plastic (polystyrene). The attenuation in these materials was measured by the pulsing method at frequencies of 1 and 3 Mc (the values most widely used in ultrasonic flaw detection) with a pulse duration of 10 μsec. The samples were rectangular bars and slabs. Longitudinal and transverse waves were exicted and received by X- and Y-cut quartz plates situated on parallel planes of the bar or slab. Rayleigh waves were excited and received in the Dural and glass by the wedge method, in the polystyrene by the comb-structure method. Acoustic coupling between the Y-cut quartz plates and surfaces of the samples was realized by means of a thin film of epoxy resin, in all other cases by means of castor oil. The decay with distance of the signal amplitude at the receiver due to divergence of the wave beam was taken into account in the calculation of all the attenuation factors from the experimental data. For the calculation of the volume wave attenuation factors according to the amplitudes of subsequent signals at the receiver corresponding to single, triple, and more traversals of the pulses over the length of the sample, the reduction in signal amplitude due to the presence of the emitting and receiving quartz plates at the reflecting surfaces was also taken into account. The relative error in determining the attenuation factors by the described procedure amounts to approximately 20%.

Also measured in conjunction with the attenuation measurements were the velocities c_l, c_t in the investigated materials. The velocities were measured by the pulsing method, using a liquid delay line with smooth regulation of the delay time used as a reference base for the pulse propagation time in the sample. The velocities c_l, c_t were used to calculate the Poisson ratio ν, then the surface wave velocity c_R.

The results of the measurements are summarized in Table 3. The factors α, β, γ were calculated from the experimentally measured values of the attenuation factors for longitudinal (δ_l), transverse (δ_t) and Rayleigh (δ_R) modes according to the relations $\alpha = \delta_l \lambda_l$, $\beta = \delta_t \lambda_t$, $\gamma = \delta_R \lambda_R$. The coefficient γ_{theor} was calculated on the basis of the values of α and β according to Eq. (I.22). As evident from the table, the relative discrepancy between the experimental and theoretical values of γ amounts to 15-20% on the average, which well supports the relation between the attenuation factors for volume and Rayleigh surface waves as described by Eq. (I.22).

§ 4. Rayleigh Waves on Cylindrical and Spherical Surfaces

Rayleigh waves can propagate not only on a plane surface, but on a cylindrical surface as well. In monograph [19], the influence of cylindrical surface curvature on the phase velocity of a plane harmonic Rayleigh wave is investigated. The characteristic equation is derived and curves are given for the dependence of the phase velocity on the curvature of the surface for two values of the Poisson ratio. In [20], the propagation of Rayleigh waves on convex and concave cylindrical surfaces is investigated in detail. Here we outline the analytical scheme and principal results of this investigation.

A convex and concave cylindrical surface are interpreted here to mean surfaces of an infinite circular cylinder and cylindrical cavity of circular cross

TABLE 3

	Material	Dural	Glass	Polystyrene
	c_l, 10^5 cm/sec	6.02	5.31	2.28
	c_t, 10^5 cm/sec	3.00	2.60	1.05
	c_R, 10^5 cm/sec	2.90	2.43	0.984
	ν	0.335	0.342	0.365
At 1 Mc	δ_l, N/cm	0.0034	0.0050	0.028
	δ_t, N/cm	0.0055	0.020	0.15
	δ_R, N/cm	0.0049	0.023	0.14
	$\alpha \cdot 10^3$	2.05	2.65	6.46
	$\beta \cdot 10^3$	1.65	5.10	15.5
	$\gamma \cdot 10^3$	1.42	5.70	13.3
	$\gamma_{theor} \cdot 10^3$	1.67	4.90	14.9
At 3 Mc	δ_l, N/cm	0.0061	0.023	0.077
	δ_t, N/cm	0.014	0.043	0.28
	δ_R, N/cm	0.018	0.046	0.22
	$\alpha \cdot 10^3$	1.22	4.02	5.80
	$\beta \cdot 10^3$	1.40	3.70	9.77
	$\gamma \cdot 10^3$	1.77	3.73	12.8
	$\gamma_{theor} \cdot 10^3$	1.40	3.72	9.50

section in an infinite elastic medium. In both cases we restrict our discussion to the plane problem, when in cylindrical coordinates r, θ, z the field in the elastic medium does not depend on z, and we will consider steady-state harmonic vibrations, when the time dependence of the field is given by the factor $e^{-i\omega t}$. The analog of the Rayleigh waves in this case may be thought of as that solution of the elasticity equations which has the following properties:

1) satisfies the conditions of zero stress on the cylindrical surface;

2) depends on the angular coordinate θ according to the law $e^{+ip\theta}$, where p is some dimensionless quantity, which may be called the angular wave number;

3) for a radius of curvature R of the cylindrical surface tending to infinity and for a finite ratio p/R, goes over to an ordinary Rayleigh wave traveling along the plane boundary between the elastic half-space and vacuum.

To remove the artificial limitation of the problem to the case of integer values of p, we apply the method of Malyuzhinets [15] for analysis of the solution in the infinite angular interval $-\infty<\theta<+\infty$, considering the axis $r=0$ to be a branch line of infinite order. In this case, for the problem of a solid cylinder we impose on the solution the added requirement of boundedness at $r=0$. With this approach the various values of the solution in the intervals $n<\theta/2\pi<n+1$ for different integers n are logically interpreted as the fields corresponding to successive traversals of the wave around the cylinder.

The solutions of both problems are expressed in terms of the potentials ϕ and ψ of longitudinal and shear waves satisfying the equations

$$\begin{aligned}\frac{1}{r}\frac{\partial}{\partial r}\left(r\frac{\partial\varphi}{\partial r}\right)+\frac{1}{r^2}\frac{\partial^2\varphi}{\partial\theta^2}+k_l^2\varphi&=0;\\ \frac{1}{r}\frac{\partial}{\partial r}\left(r\frac{\partial\psi}{\partial r}\right)+\frac{1}{r^2}\frac{\partial^2\psi}{\partial\theta^2}+k_t^2\psi&=0.\end{aligned}\qquad(\text{I}.23)$$

The displacement components U_r, U_θ and stress components σ_{rr}, $\sigma_{r\theta}$ may be represented in terms of ϕ and ψ according to the equations

$$\begin{aligned}U_r&=\frac{\partial\varphi}{\partial r}+\frac{1}{r}\frac{\partial\psi}{\partial\theta},\\ U_\theta&=\frac{1}{r}\frac{\partial\varphi}{\partial\theta}-\frac{\partial\psi}{\partial r};\end{aligned}\qquad(\text{I}.24)$$

$$\begin{aligned}\sigma_{rr}&-(\lambda+2\mu)\left(\frac{\partial^2\varphi}{\partial r^2}-\frac{1}{r^2}\frac{\partial\psi}{\partial r}+\frac{1}{r}\frac{\partial^2\psi}{\partial r\partial\theta}\right)+\\ &+\lambda\left(\frac{1}{r^2}\frac{\partial^2\varphi}{\partial\theta^2}-\frac{1}{r}\frac{\partial^2\psi}{\partial r\partial\theta}+\frac{1}{r}\frac{\partial\varphi}{\partial r}+\frac{1}{r^2}\frac{\partial\psi}{\partial\theta}\right),\\ \sigma_{r\theta}&=\mu\left(\frac{2}{r}\frac{\partial^2\varphi}{\partial r\partial\theta}+\frac{1}{r^2}\frac{\partial^2\psi}{\partial\theta^2}-\frac{\partial^2\psi}{\partial r^2}-\frac{2}{r^2}\frac{\partial\varphi}{\partial\theta}+\frac{1}{r}\frac{\partial\psi}{\partial\theta}\right).\end{aligned}\qquad(\text{I}.25)$$

Convex Cylindrical Surface. The only solutions of Eqs. (I.23) depending on θ solely in terms of the factor $e^{ip\theta}$ and bounded at $r=0$ are the following:

$$\begin{aligned}\varphi&=Ae^{ip\theta}I_p(k_lr);\\ \psi&=Be^{ip\theta}I_p(k_tr),\end{aligned}\qquad(\text{I}.26)$$

where A and B are arbitrary constants, $I_p(k_l r)$, $I_p(k_t r)$ are Bessel functions of order p. The angular wave number p is equal to $2\pi R/\lambda$, R is the radius of the cylinder, λ is the spatial period on the surface of the cylinder (not to be confused with the elastic Lamé constant). Henceforth we will call the solution a surface wave (as opposed to a Rayleigh wave propagating on a plane surface), λ the surface wavelength, $k = 2\pi/\lambda$ the wave number, amd $c = \omega/k$ the phase velocity of this wave. Inasmuch as we are considering a wave traveling along θ, p may be either integer or fractional.

The condition that the stresses σ_{rr} and $\sigma_{r\theta}$ be equal to zero at $r = R$ yields two equations, taking (I.26) into account. Making use of the identity relations between Bessel functions, we write these equations as

$$\frac{B}{A} = \frac{I_{p-2}(x) - I_{p+2}(x)}{I_{p-2}(y) + I_{p+2}(y)} \cdot \frac{k_l^2}{k_t^2} i; \tag{I.27}$$

$$\frac{I_{p+2}(x) + I_{p-2}(x) - 2\left(\frac{k_t^2}{k_l^2} - 1\right) I_p(x)}{I_{p+2}(x) - I_{p-2}(x)} = \frac{I_{p+2}(y) - I_{p-2}(y)}{I_{p+2}(y) + I_{p-2}(y)}, \tag{I.28}$$

where $x = k_l R$, $y = k_t R$.

The first of these equations gives the amplitude relation between ϕ and ψ for known p. The second equation is the characteristic equation for the given case and defines the relationship of p with x and y, i.e., it defines the wave number k for a given cylinder radius R. One is readily convinced that for a given cylinder radius R, i.e., for given x and y, there exist many values of p that will satisfy Eq. (I.28). Each of these values of p determines the phase velocity of the corresponding wave. We will investigate the wave which is localized near the surface and which goes over to a Rayleigh wave when the radius of curvature of the surface is increased to infinity. Such a wave corresponds to just one root of Eq. (I.28), and conversely. We will determine the solution of Eq. (I.28) only for this root.

It is convenient to solve this equation graphically, specifying the value of p and finding the point of intersection of the curves corresponding to the left- and the right-hand sides of the equation. The results of the solution are shown in Fig. 14 for media with Poisson ratios $\nu = 0.25$ (curve 1) and $\nu = 0.34$ (curve 2). The phase velocity ratios of the surface waves to Rayleigh waves in the corresponding media are plotted on the ordinate axis, the values of $p = 2\pi R/\lambda$ on the abscissa axis. As evident from the graph, the ratios c/c_R for

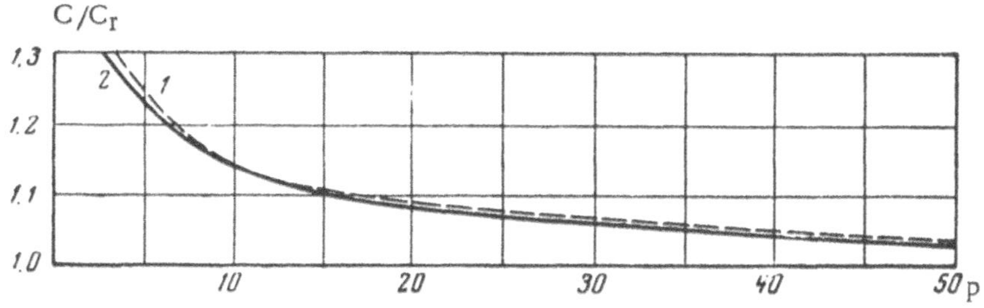

Fig. 14

these media, given the same value of p, are practically identical. Moreover, it is also apparent from the graph that $c > c_R$ and $c \to c_R$ as $p \to \infty$. The latter assertion is indicative of the fact that the surface wave goes over to a Rayleigh wave as $p \to \infty$. This is easily verified by taking the asymptotic representations of $I_p(k_l r)$, $I_p(k_t r)$, $I_p(k_l R)$, $I_p(k_t R)$ as $p \to \infty$ and $(R-r)/r \to 0$ [22]. Equation (I.28) then goes over to the ordinary Rayleigh equation, and the expressions (I.26) go over to the analogous expressions for a plane Rayleigh wave traveling over the boundary of an elastic half-space with a vacuum.

The investigated surface wave is similar to a Rayleigh wave. Its phase velocity c is near c_R, and the displacements are concentrated in a thin surface layer of the order of wavelength in thickness. The displacements are calculated from Eqs. (I.24) with regard for (I.27) and (I.28). The dependence of the displacement amplitudes on the depth for a medium with Poisson ratio 0.25 is illustrated in Fig. 15 for three values of p: $p = 5$, $p = 41$, $p = \infty$ (the latter case corresponding, of course, to a Rayleigh wave and $\hat{U}_{r0}$ denoting in the figure the amplitude of the radial displacement in the surface wave at $r = R$). The dependences cited show that the displacements in the surface wave decay somewhat more rapidly with distance from the surface than in a Rayleigh wave, the rate of decay being faster for smaller $p = kR$. At $r = 0$, the displacements in the surface wave are equal to zero, whereas in a Rayleigh wave the displacements do not vanish altogether at any depth. Qualitatively, the cited dependences of the displacements on the depth tend to remain the same for any elastic medium.

For large values of p (of the order of 100 or more) the following expression is valid for the surface wave phase velocity:

$$c = c_R(1 + \delta), \tag{I.29}$$

where δ is a small correction depending on the elastic properties of the

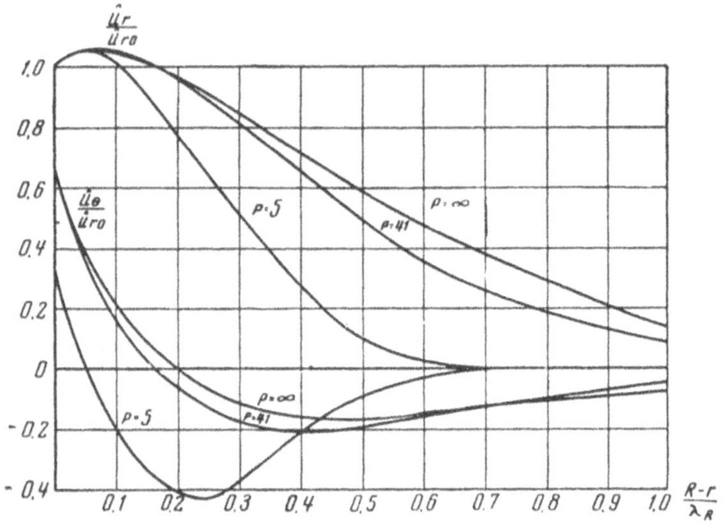

Fig. 15

medium and p, where $\delta \to 0$ as $p \to \infty$. It is conveniently determined by rewriting Eq. (I.28) in the form

$$\frac{I_{p-2}(x)}{I_p(y)} \cdot \frac{I_{p+2}(y)}{I_p(y)} + \frac{I_{p+2}(x)}{I_p(x)} \cdot \frac{I_{p-2}(y)}{I_p(y)} - \\ - \left(\frac{k_t^2}{k_l^2} - 1\right)\left[\frac{I_{p+2}(y)}{I_p(y)} + \frac{I_{p-2}(y)}{I_p(y)}\right] = 0. \tag{I.30}$$

For large values of the argument (x and y), as well as the index (p), and with the stipulation that the argument is smaller than the index ($k_l R$, $k_t R < kR$), which is always true in the given case since $c_R < c_t < c_l$, the following asymptotic representation in terms of a semiconvergent Debye series [22] is valid for the Bessel functions:

$$I_p(\gamma) = \frac{e^{p(\operatorname{th}\alpha - \alpha)}}{\sqrt{2\pi p \operatorname{th}\alpha}}\left[1 + \frac{1}{p}\left(\frac{1}{8}\operatorname{cth}\alpha - \frac{5}{24}\operatorname{cth}^3\alpha\right) + \ldots\right], \tag{I.31}$$

where $p/\gamma = \cosh\alpha$. Making use of this representation, we find an approximate expression for the left-hand side of Eq. (I.3) correct to terms of order $1/p^2$. Equating it to zero, we obtain an approximate frequency equation. The leading terms therein are of order unity; they form an equation for the determination of c_R. The remaining terms yield an equation for the determination of δ, from which we obtain

$$\delta = \frac{\left[1+\frac{s_R}{q_R}+\frac{k_R}{2s_R}\left(1-\frac{s_R^2}{q_R^2}\right)\right] e^{2\operatorname{arcth}\frac{q_R}{k_R}-2\operatorname{arcth}\frac{s_R}{k_R}}}{D} +$$

$$+\frac{\left[1+\frac{s_R}{q_R}-\frac{k_R}{2s_R}\left(1-\frac{s_R^2}{q_R^2}\right)\right] e^{-2\operatorname{arcth}\frac{q_R}{k_R}+2\operatorname{arcth}\frac{s_R}{k_R}}-\frac{2s_R^2}{k_t^2}\left(\frac{k_t^2}{k_l^2}-1\right)}{D}, \quad \text{(I.32)}$$

$$D = k_R R\left[\left(1-\frac{s_R}{q_R}\right)\left(e^{2\operatorname{arcth}\frac{q_R}{k_R}-2\operatorname{arcth}\frac{s_R}{k_R}}-\right.\right.$$

$$\left.\left.-e^{-2\operatorname{arcth}\frac{qR}{k_R}+2\operatorname{arcth}\frac{sR}{k_R}}\right)+\frac{4k_R s_R}{k_t^2}\left(\frac{k_t^2}{k_l^2}-1\right)\right].$$

Consequently, the phase velocity $c = c_R(1+\delta)$ for a surface wave on a convex cylindrical surface has a correction δ (by comparison with c_R) proportional to $1/k_R R$. We see at once that $\delta > 0$ for any elastic medium.

The group velocity may be found from the familiar relation between the phase and group velocities. It turns out in the given approximation for the phase velocity that $c^{gr} = c_R$, i.e., the group velocity has a correction of the order $(1/k_R R)^2$.

Concave Cylindrical Surface. In the present instance the only solutions of (I.23) depending on θ according to the law $e^{ip\theta}$ and satisfying the extinction principle [15] over the entire region $r \geq R$ are the following:

$$\varphi = Ae^{ip\theta}H_p^{(1)}(k_l r); \qquad \text{(I.33)}$$
$$\psi = Be^{ip\theta}H_p^{(1)}(k_t r).$$

Here, A and B are arbitrary constants, $H_p^{(1)}(k_l r)$, $H_p^{(1)}(k_t r)$ are p-th-order Hankel functions of the first kind. Equations (I.27) and (I.28) are replaced by their analogies:

$$\frac{B}{A} = \frac{H_{p-2}^{(1)}(x) - H_{p+2}^{(1)}(x)}{H_{p-2}^{(1)}(y) + H_{p+2}^{(1)}(y)} \cdot \frac{k_l^2}{k_t^2}\, i, \qquad \text{(I.34)}$$

$$\frac{H^{(1)}_{p+2}(x) + H^{(1)}_{p-2}(x) - 2\left(\frac{k_t^2}{k_l^2} - 1\right) H^{(1)}_p(x)}{H^{(1)}_{p+2}(x) - H^{(1)}_{p-2}(x)} = \frac{H^{(1)}_{p+2}(y) - H^{(1)}_{p-2}(y)}{H^{(1)}_{p+2}(y) + H^{(1)}_{p-2}(y)}. \quad \text{(I.35)}$$

The latter equation is the frequency equation for the given case. We will investigate its solutions corresponding to a wave localized near the surface. It is clear that even for real x and y this equation is only satisfied for complex $p = p_1 + ip_2$. The wave number in this case is also complex:

$$k = k_1 + ik_2. \quad \text{(I.36)}$$

Since $p = kR$, we have

$$p_1 = k_1 R, \; p_2 = k_2 R. \quad \text{(I.37)}$$

Consequently, a surface wave on a concave cylindrical surface will propagate with attenuation. This attenuation is associated with the radial radiation of energy into the medium, in this case by the surface wave. In fact, for $r \gg R$ the functions $H_p^{(1)}(k_l r)$ and $H_p^{(1)}(k_t r)$ from (I.33) represent cylindrical waves departing in the radial direction.

We replace the Hankel functions in Eq. (I.35) by their expressions in terms of the Bessel functions $I_p(x)$, $I_p(y)$ and Neumann functions $N_p(x)$, $N_p(y)$ of the same order.

For large values of p ($p > 100$) we use the asymptotic representation (I.31) for the Bessel functions $I_p(x)$ and $I_p(y)$, for the Neumann functions we take the analogous representation from [22]. It can then be shown that as $k_1R \to \infty$, Eq. (I.35) goes over to the Rayleigh equation, and the expressions (I.33) transform to the corresponding expressions for a Rayleigh wave, i.e., the surface wave goes over to a Rayleigh wave for $k_1R \to \infty$, as in the case of a convex surface for $kR \to \infty$.

For large but still finite k_1R the following representation holds for the surface wave number:

$$k = k_R(1 - \zeta) + ik_2. \quad \text{(I.38)}$$

Here, k_2 is the imaginary part of the wave number; $k_2 \to 0$ as $k_1R \to \infty$; ζ is a small correction depending on the elastic properties of the medium and the values of k_1R, and vanishes for $k_1R \to \infty$. Accordingly, for the phase velocity of the wave we obtain the expression

$$c = c_R(1+\zeta). \tag{I.39}$$

Making use of the asymptotic representations for the Bessel and Neumann functions and assuming that k_1R is no greater than unity in order of magnitude (as we will see below, this assumption is correct), we obtain the following expressions for ζ and k_2 from the real amd imaginary parts of Eq. (I.35), after a series of transformations:

$$\zeta = -\delta,$$

$$k_2 = \frac{q_R s_R (k_R^2 + s_R^2)^2}{4k_R(q_R - s_R)[k_R^2(q_R - s_R) + 2q_R s_R^2]} e^{-2k_1R\left(\operatorname{arcth}\frac{s_1}{k_1} - \frac{s_1}{k_1}\right)}, \tag{I.40}$$

where

$$s_1 = \sqrt{k_1^2 - k_t^2}. \tag{I.41}$$

Comparing (I.40) and (I.32), we note that the corrections to the phase velocities of surface waves on a convex and a concave cylindrical surface of the same radius R differ only in sign (hence, always $\zeta < 0$). The group velocity of the waves on the concave cylindrical surface, as in the case of the convex cylindrical surface, is equal to c_R correct to terms of order $(1/k_RR)^2$. It follows from Eq. (I.41) that the assumption of small k_2R is justified; the imaginary part of the wave number k_2 is small for large values of $k_1R(k_2 \sim e^{-2k_1R})$.

Applying Eq. (I.41), it is not too difficult to calculate the coefficient γ of excess absorption due to surface wave radiation into the depth over the Rayleigh wavelength on a concave cylindrical surface, according to the relation $\gamma = k_2\lambda_R = 2\pi(k_2/k_R)$. For $k_RR \simeq 100$ $(R/\lambda_R \simeq 20)$ the relative error in calculating γ and k_2 is slight; it is of the order of magnitude $1/(k_RR)^2$. For $k_RR \simeq 30$ $(R/\lambda_R \simeq 5)$ the relative error does not exceed 30%. Consequently, Eq. (I.41) may be used for approximate calculations of γ and k_2 for relatively large values of k_RR. The dependence of $\ln \gamma$ on R/λ_R is illustrated in Fig. 16. As apparent from that figure, the excess attenuation of surface waves on a concave cylindrical surface is strongly dependent on the curvature of the surface. For $R/\lambda_R \simeq 5$ $(k_RR \simeq 30)$ the attenuation is very high, for $R/\lambda_R \simeq 50$ $(k_RR \simeq 300)$ it is insignificantly small. For the sake of illustration we present below some values of the quantity $e^{-(\pi R/\lambda_R)\gamma}$ for several radii of curvature of the cylindrical surface, characterizing the surface wave attenuation on a concave cylindrical surface over a path equal to the semicircumference associated with the radius R (at the beginning of the path the wave amplitude is assumed equal to unity):

R/λ_R*	5	10	15	20	30	50
$e^{-(\pi R/\lambda_R)\gamma}$	0.280	0.340	0.555	0.751	0.948	0.999

*The cited data refer to the case $\nu = 0.34$.

In [23, 24], which we review in more detail below, experiments are described in which it is possible to determine the values of the phase velocities for surface waves on convex and concave cylindrical surfaces of fairly small radius of curvature ($R/\lambda_R \sim 1-2$). The resultant experimental values of the phase velocities are in good agreement with the theoretical.

An experimental investigation of the propagation of Rayleigh-type surface waves on cylindrical surfaces is described in [21], the objective being in this case an experimental verification of the theoretical results outlined above with respect to the attenuation of such waves on convex and concave cylindrical surfaces. The experiments were performed in the pulsed mode at a frequency of 2.65 Mc and pulse durations of 10 μsec (concave cylindrical surface) and 5 μsec (convex cylindrical surface). The lateral surfaces of Dural disks with radii R = 5−85 mm and a thickness of 25 mm and the surfaces of semicylindrical recesses with radii 1−50 mm cut in the laterial surfaces of Dural bars with a 25 × 70 mm rectangular cross section were used for the convex and concave cylindrical surfaces for the attenuation experiments (see Fig. 17). As a plane surface for attenuation measurements, the lateral surface of a Dural bar of the same rectangular cross section was used (control bar). The ends of the control bar (Fig. 17) were beveled out of perpendicular, with the axis in the form of a dihedral end face (more concerning this below). Since the value of the Rayleigh wave attenuation factor depends considerably

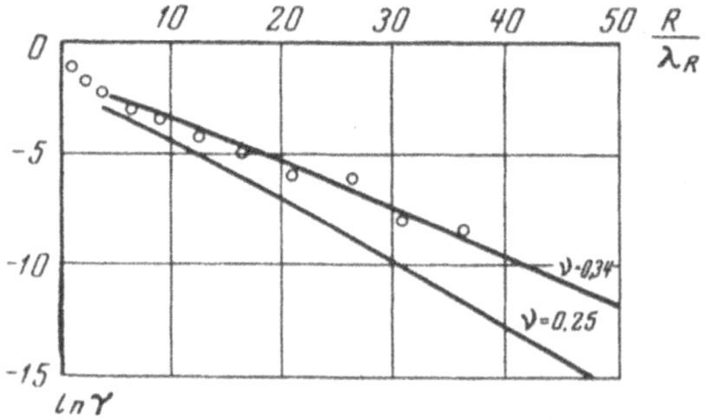

Fig. 16

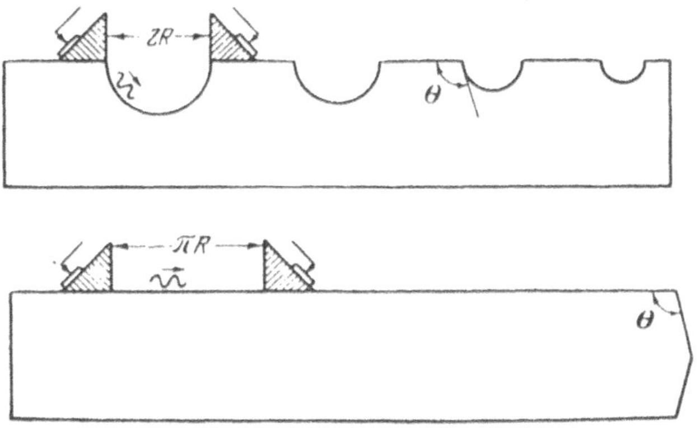

Fig. 17

on the structure of the material and the degree of surface finish, in order to preclude the influence of these factors all of the disks and bars were made from the same sheet of Dural D16, and the plane and cylindrical surfaces were machined and finished with strict uniformity.

The wave attenuation on the concave cylindrical surface was determined as follows (see Fig. 17). The pulse amplitudes were compared at the indicator with the prismatic emitter and receiver placed on the edges of the recess of radius R (amplitude A_1) and on the surface of the control bar at a distance πR (amplitude A_2). The ratio A_2/A_1, divided by the product of the Rayleigh wave transmission coefficients for propagation from the plane surface of the bar to the cylindrical surface of the recess (K') and from the cylindrical surface of the recess to the surface of the bar (K"), is equal to $e^{-(\pi R/\lambda_R \gamma)}$, where γ is the coefficient of excess attenuation along the path λ_R on the cylindrical surface, over and above that on the plane surface. It was assumed in calculating γ that K' = K". The coefficient K' was determined experimentally as the coefficient of Rayleigh wave transmission from one face of the elastic wedge with corner angle θ to the other face, where $\theta \sim 90°$ is the dihedral angle between the plane surface of the bar and the tangential plane drawn to the surface of the recess at half the depth of the Rayleigh-wave localization layer (see Fig. 17). Analogous measurements were made on the lateral and end surfaces of the control bar. For small R ($R \sim \lambda_R$), a correction is introduced into the coefficient K' on the basis of data from [24] to account for the transformation of Rayleigh waves at the plane surface into surface Rayleigh waves on the cylindrical surface, which differ appreciably from the former when $R \sim \lambda_R$. In order to eliminate the unstable influence of the oil transition layer between the surfaces of the emitting–receiving prisms and the bar on the measurements of A_1 and A_2, these measurements were repeated 20 times, after which they were averaged. As a result, the maximum random error introduced by acoustic coupling instability in measuring the amplitudes A_1 and A_2 was reduced from 40% (for single measurements) to 20%, a wholly acceptable figure. The potential on the emitter was monitored and held constant. The same averaging was used in the experiments with convex cylindrical surfaces. The total error in determining γ by the method described did not exceed 15% for $R/\lambda_R \sim 10$ and amounted to $\sim$30% for $R/\lambda_R \sim 40$, when the absolute value of γ became very small. This error was attributable to instrument error and the random error of measurement. An additional cause of error might have been a certain possible fluctuation in the absorption and scattering of surface Rayleigh waves on the concave cylindrical surface (by comparison with the plane surface) elicited not so much by the geometry of the surface as by a certain discrepancy in the velocity and structure of the Rayleigh waves on the plane and cylindrical surfaces. We note that an error source of this type could also have occurred in the experiments described below with convex cylindrical surfaces.

The experimental values of γ are represented by dots in Fig. 16. As evident from the figure, these values are in satisfactory agreement with the corresponding theoretical values (curve for $\nu = 0.34$). This agreement confirms the validity of the approximate analytical equation (I.41) over a wide interval of $k_R R$ ($k_R R \gtrsim 30$, $R/\lambda_R \gtrsim 5$).

The purpose of the experimental investigations of surface-wave attenuation on convex cylindrical surfaces was to demonstrate the lack of excess attenuation of surface Rayleigh waves on these surfaces over and above the plane surface. Toward this end, the amplitude attenuation of a surface-wave pulse with distance from the emitter was measured for propagation over convex cylindrical surfaces of various radii, including $R = \infty$ (plane surface). The emitter in these experiments was a square titanate plate with dimensions 9×9 mm and natural resonant frequency of 2.5 Mc, the plate being mounted in a fixed position on the indicated surfaces and acoustically coupled to them with a film of castor oil. For small R ($R \sim 5-10\lambda_R$), this plate also served as a surface wave receiver, which measured a series of successive values of the surface-wave pulse amplitudes corresponding to successive traversals of the pulse around the periphery of the disk. The desired amplitude decay of the pulse with distance from the source could easily be measured from the indicated series of amplitude values, taking into account the pulse attenuation factor due to its transmission through the region of coupling between the plate and disk.

The attenuation factor was determined experimentally. For this purpose, a "prismatic" surface wave emitter and receiver were placed on the

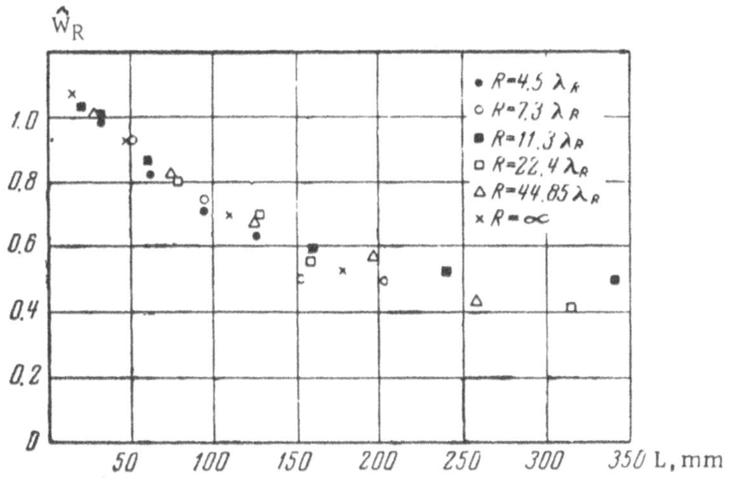

Fig. 18

cylindrical surface of the disk on either side of the titanate plate, and the pulse amplitude at the receiver was compared with, and without, the titanate plate present between the emitter and receiver. For large R(R ~ 40λ) and in the case of a plane surface, the amplitude attenuation of the surface wave pulse with distance from the source was measured directly by a "prismatic" receiver placed at various distances from the titanate emitter plate. For medium R(R ~ 20λ), the amplitude attenuation was measured by both of the techniques described. The overall error of the measurements in the experiments with convex cylindrical surfaces was at most 10-15% for all R values.

The results of the measurements are shown in Fig. 18. The distance L from the emitter in millimeters, measured along the cylindrical surface perpendicular to the generatrices, is plotted on the horizontal axis; the amplitude of the surface Rayleigh wave pulse on the cylindrical surfaces of various radii (including R = ∞) is plotted on the vertical axis in arbitrary units. The amplitudes referring to a surface of the same radius are denoted by identical symbols. All of the amplitudes have been normalized; their values at L = 30 mm are assumed equal to unity (the values of the amplitudes at the point L = 30 mm prior to normalization were computed from the two adjacent values of the corresponding amplitudes by means of linear interpolation). It is clear from Fig. 18 that, within the limits of experimental error, all of the experimental points fit a single curve. This demonstrates the independence of the attenuation factor for surface Rayleigh waves on a convex surface from the radius of curvature.

Spherical Surface. Waves of the Rayleigh type can also exist on a spherical surface. The problem of harmonic waves of this type on the surface of a perfectly elastic sphere of radius R has been treated in [25]. Rayleigh-type waves are interpreted as the exact solution of the elasticity equations which satisfies the condition of zero stresses on the surface of the sphere and which has the character of steady-state monochromatic surface waves. At the poles of sphere $\theta = 0$ and $\theta = \pi$ (in spherical coordinates r, ϕ, θ) were placed a wave "source" and "sink," corresponding to the singular points of the solutions. It was assumed that the "source" and "sink" were completely equivalent to one another and that waves propagated from the poles with equal amplitudes in the $+\theta$ and $-\theta$ directions, so that their addition would result in the formation of standing waves which were regular at all points of the sphere.

With the problem thus stated, the asymptotic expression for the particle displacement vector **u** in the wave for a sphere of radius R much greater than the surface wavelength has the following form, given $R - r \ll R$ (near the surface of the sphere) and $0 < \theta < \pi$:

$$\mathbf{u} = \sqrt{\frac{1}{\sin\theta}}\left[e^{-\frac{m}{R}\sqrt{1-k_l^2/k_R^2}(R-r)}\left(-\mathbf{r}_0\sqrt{1-k_l^2/k_R^2}\cos\Phi + \right.\right.$$
$$\left. + \boldsymbol{\theta}_0 \sin\Phi\right) + \frac{2\sqrt{1-k_l^2/k_R^2}}{2-k_l^2/k_R^2}\, e^{-\frac{m}{R}\sqrt{1-k_t^2/k_R^2}(R-r)}\left(\mathbf{r}_0\cos\Phi - \right. \tag{I.42}$$
$$\left.\left. -\boldsymbol{\theta}_0\sqrt{1-k_t^2/k_R^2}\sin\Phi\right)\right].$$

Here $\mathbf{r}_0$, $\boldsymbol{\theta}_0$ are unit vectors in the spherical coordinate system; m is any integer much greater than one; $\Phi = m/R(\theta R \pm c_R t)$ is the phase multiplier.

It is readily seen from Eq. (I.42) that, as in the case of a plane surface, the amplitude of the displacement vector decays exponentially with depth, and the phase velocity coincides with the velocity on a plane surface. The influence of a small spherical curvature ($R \gg \lambda_R$) only leads to a discrete frequency spectrum on the part of the steady-state waves. In fact, for waves on a sphere, the role of the wave number $k_R = \omega/c_R$ is taken by the ratio m/R, which assumes only discrete values (this being a result of the condition stipulated in the statement of the problem regarding the equivalence of the "source" and "sink"). For $R \to \infty$, the frequency spectrum goes over to a continuous spectrum.

§ 5. Rayleigh Waves on Dihedral Surfaces

If two plane surfaces form a dihedral angle, a Rayleigh wave propagating over one of the faces will, on arriving at the apical edge, be partially reflected and partially transmitted to the second surface. An experimental investigation of this phenomenon on Dural samples in the form of bars with dihedral apex angles between the limits of 12 and 170° is described in [23].

A rectangular pulse of Rayleigh waves with a duration of 10 μsec and carrier frequency of 2.70 Mc was propagated over one of the faces of the dihedron (wedge) perpendicular to the edge (Fig. 19). The coefficients of reflection from the edge and transmission of the pulse to the other face were measured. The coefficients were measured in terms of the amplitudes of the corresponding pulses. The reflection coefficient K_{re} was measured as follows. The amplitude A_1 of the pulse traveling from the emitter was measured at a distance l_1 before reaching the edge of the wedge. Then the amplitude A_2 of a pulse reflected from the edge was measured at the same point, using the same receiver. The ratio A_2/A_1, multiplied by a coefficient accounting for the attenuation of the first pulse as a result of its transmission through the portion of the surface taken up by the receiving unit, as well as the divergence and absorption of the wave beam over the path $2l_1$, is the desired reflection coefficient. The transmission coefficient K_{tr} was defined as the ratio of the

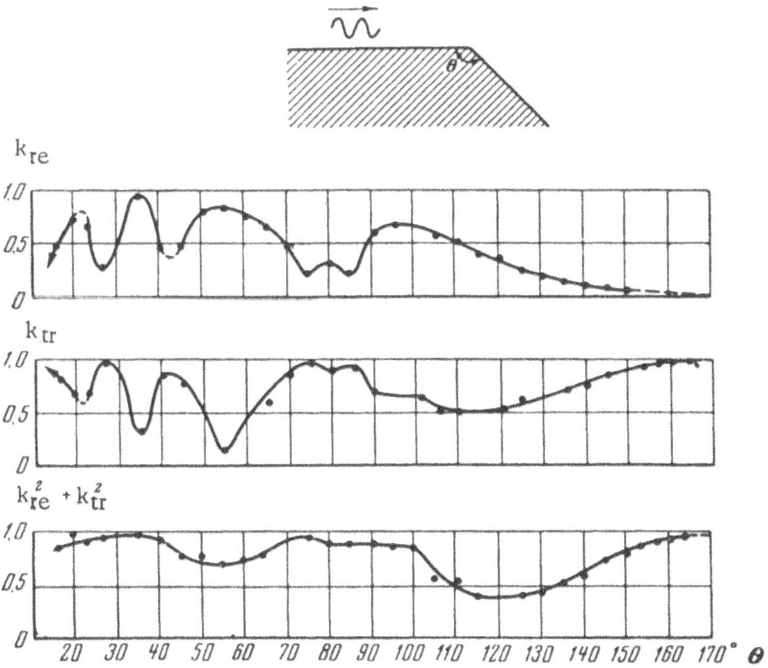

Fig. 19

amplitude of a pulse at a distance l_2 beyond the edge of the wedge (on the second face) to the amplitude A_1 of a pulse at a distance l_1 ahead of the edge, multiplied by a coefficient accounting for the amplitude attenuation of the wave beam over the path $l_1 + l_2$.

The results of the measurements are shown in Fig. 19. The apex angle of the wedge is plotted on the horizontal axis in degrees, the coefficients K_{re}, K_{tr}, and the sum of their squares, representating the ratio of the total energy of the transmitted and reflected Rayleigh waves to the energy of the incident wave, are plotted on the vertical. As apparent from the graph, the sum $K_{re}^2 + K_{tr}^2$ is always less than one. This attests to the constant conversion of a part of the energy of the incident Rayleigh wave into the energy of longitudinal and transverse waves scattered by the faces and edge of the wedge into the depth of the medium. The transformation of the waves is caused by the fact that the aggregate of the incident, reflected, and transmitted Rayleigh waves does not satisfy the condition of zero stress on the faces of the wedge. It is also evident from the graphs that the reflectivity and transmittivity of the wedge depend strongly on the apex angle θ; the curves for $K_{re}(\theta)$ and $k_{tr}(\theta)$ have clearly pronounced maxima and minima, the maxima of the reflection coefficient, as a rule, corresponding to minima of the transmission coefficient, and vice-versa (except for the case $\theta = 115°$). As θ approaches 180°, the coefficient $K_{re} \to 0$, while $k_{tr} \to 1$. The transmission and reflection coefficients never reach the values of zero or one. Curves are given in [17] for $K_{re}(\theta)$, $k_{tr}(\theta)$, $K_{re}^2(\theta) + K_{tr}^2(\theta)$ in the case of a steel wedge. The curves are similar to those described above. This strongly suggests that the behavior of the curves will be the same for any elastic medium.

An experimental investigation of the transmission and reflection of Rayleigh waves at the faces of a wedge was also carried out in [26], wherein results similar to those cited above were obtained. In [26], however, the investigation pertained to seismology problems, rather than to ultrasonic measurements.

If the edge between the faces of the wedge is rounded, the transmission of Rayleigh waves from one face to the other is greatly improved. The results of experiments on the transmission and reflection of Rayleigh waves by cylindrical rounded edges of radius 0 to $1.7\lambda_R$, made between the lateral and end surfaces of rectangular aluminum bars (Fig. 20), are described in [24]. The case of normal incidence of a Rayleigh wave on a rounded edge was analyzed. The measurements were carried out according to the same scheme as in [23].

The results of the measurements are shown in Fig. 20. As apparent from the figure, as the radius of the rounded edge is increased, the coefficient K_{tr} increases, tending to unity, while the coefficient K_{re} decreases, tending to

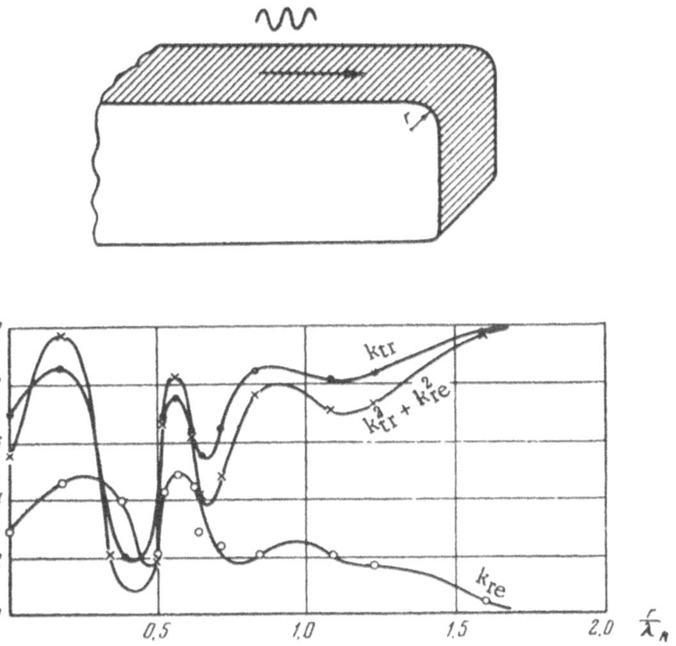

Fig. 20

zero. However, the coefficients K_{tr} and K_{re} increase and decrease in a strongly oscillating, rather than a monotonic, course. It may be assumed by analogy with the transmission and reflection of waves in plane layers that the oscillations are attributable to the interference mechanism behind the formation of the transmitted and reflected Rayleigh waves. The reflected wave is formed as the result of interference between reflections from the leading and back edges of the rounded edge. The transmitted Rayleigh wave arises in analogous fashion. The phase difference between the indicated reflections is determined by the number of half-waves that fit into the rounded arc. These waves are the analog of Rayleigh waves on a convex cylindrical rounded surface. Their phase velocity c_{cvx}, as shown in [20], is always greater than c_R and depends on r/λ_R. From the distance between the maxima and minima of the curves for $K_{tr}(r/\lambda_R)$ and $K_{re}(r/\lambda_R)$ in the interval $0.20 < r/\lambda_R < 1.15$, it is possible to ascertain the experimental value of the average (in the indicated interval) velocity c_{cvx} for aluminum; this value comes to $1.29c_R$. The corresponding theoretical value of c_{cvx}, according to [20], is equal to $1.27c_R$ and, hence, is in good agreement with the experimental, confirming the interference mechanism of the transmission and reflection of Rayleigh waves in a rounded section. The curves for $K_{tr}^2 + K_{re}^2$, as in the wedge case, pass below unity, again evincing conversion of a part of the energy of the incident Rayleigh wave into energy of longitudinal and transverse modes scattered by the rounded edge.

The results described herein foster the conclusion that for $r/\lambda_R < 1.7$ the transmission and reflection of Rayleigh waves by a rounded section are determined by the ratio r/λ_R, while for $r/\lambda_R > 1.7$ the transmission and reflection coefficients become practically equal to one and zero, respectively, i.e., "total" transmission of the Rayleigh waves through the rounded portion sets in. We point out in conclusion that, considering the mechanism of the formation of the reflected and transmitted Rayleigh waves, it is reasonable to expect that the qualitative behavior of the given dependences $K_{tr}(r/\lambda_R)$ and $K_{re}(r/\lambda_R)$ will be preserved for any elastic medium and for any apex angles of a wedge with its edge rounded off.

§ 6. Propagation at an Interface with a Liquid

At the boundary between a solid half-space and a liquid layer of finite or infinite thickness it is possible for surfaces waves similar to the Rayleigh type to exist.

We will first consider the problem of the propagation of plane harmonic surface waves at the boundary between two half-spaces – a solid and a liquid. We will assume that propagation is in the direction of the x axis, the z axis being directed perpendicular to the interface into the depth of the solid half-space (see Fig. 1). The solid half-space will be regarded as a homogeneous, isotropic, perfectly elastic medium, the liquid as an ideal liquid. The expressions for the potentials ϕ and ψ for longitudinal and transverse waves in the solid half-space must (as in the Rayleigh wave case) satisfy the wave equations (I.2), while the expressions for the potential ϕ_L in the liquid must satisfy the analogous equation

$$\frac{\partial^2\varphi_L}{\partial x^2} + \frac{\partial^2\varphi_L}{\partial z^2} + k_L^2\varphi_L = 0, \tag{I.43}$$

where $k_L = \omega/c_L$ is the wave number in the liquid (henceforth, all physical quantities pertaining to the liquid will be designated by the subscript L, those pertaining to the solid being labeled as in Section 1. Inasmuch as we seek a surface wave solution, the potentials ϕ and ψ in the solid must describe (as in the Rayleigh wave case) longitudinal and transverse inhomogeneous waves propagating in the x direction and decaying in the z direction. The potential ϕ_L must correspond in the liquid to a wave that will "fit" the two waves just mentioned. This means that this wave must pursue a path along the x axis with a velocity equal to the phase velocity of the longitudinal and transverse inhomogeneous waves. At the plane z = 0, the three indicated waves must satisfy the boundary conditions of equal normal displacement W in the liquid and solid, pressure in the liquid equal to the stress σ_{zz} in the solid, and zero tangential stress σ_{xz}.

Carrying out some simple computations (similar to those in Section 1), we obtain the following relations for the components of the displacements along the x and z axes in the liquid and solid:

$$\left.\begin{aligned} U_L &= -Ak\frac{qk_t^2}{\sqrt{k_L^2-k^2}\,(k^2+s^2)}\,e^{i\left(kx-\sqrt{k_L^2-k^2}\,z-\omega t\right)} \\ W_L &= Ak_t\frac{qk_t}{k^2+s^2}\,e^{i\left(kx-\sqrt{k_L^2-k^2}\,z-\omega t\right)}; \end{aligned}\right\} \tag{I.44}$$

$$\left.\begin{aligned} U &= Ak\left[e^{-qz}-\frac{2qs}{k^2+s^2}\,e^{-sz}\right]e^{i\left(kx-\omega t+\frac{\pi}{2}\right)}, \\ W &= -Aq\left[e^{-qz}-\frac{2k^2}{k^2+s^2}\,e^{-sz}\right]e^{i(kx-\omega t)}. \end{aligned}\right\} \tag{I.45}$$

Here, k is the surface wave number, A is an arbitrary constant. The characteristic equation determining the wave number k has the form

$$4k^2qs-(k^2+s^2)^2=i\,\frac{\rho_L}{\rho}\cdot\frac{qk_t^4}{\sqrt{k_L^2-k^2}}. \tag{I.46}$$

It is readily perceived that this equation differs from Eq. (I.6) for a Rayleigh wave in the presence of an additional term on the right-hand side to account for the effect of the liquid.

It is shown in [27] that for any relation between the parameters of the solid and liquid media Eq. (I.46) has one real root corresponding to a surface wave traveling along the boundary with a phase velocity less than the velocity c_L of the wave in the liquid and the velocities $c_{l,t}$ of the longitudinal and transverse waves in the solid. The indicated wave consists of an inhomogeneous wave in the liquid and two inhomogeneous waves in the solid. All three waves decay exponentially with distance (in both directions) from the boundary z = 0. As shown in the monograph [28], this surface wave has an altogether different structure and velocity from a Rayleigh wave (even if the density of the liquid is small in comparison with the density of the solid). The wave energy and motion of the particles are mainly localized in the liquid, rather than in the solid. In ultrasonic practice, therefore, this wave mode is not used, and we will not discuss it further.

It is shown in the monograph [28] that, given the condition $c \lessapprox c_R$, which is fulfilled for almost all real media, Eq. (I.46) has (in addition to the real root) a complex root k corresponding to a system of three waves (one in the liquid and two in the solid), which transforms to a Rayleigh wave in the solid as the density of the liquid tends to zero. The complexity of the root has a simple physical interpretation; the surface wave in this case radiates energy continuously into the liquid, forming therein an inhomogeneous wave departing from the boundary. This type of wave, which is similar to a Rayleigh wave, is one of particular interest with regard to immersion ultrasonic flaw detection and other areas of ultrasonic engineering. This type of surface wave has been investigated by a number of authors, but always with the assumption that the influence of the liquid on the solid is slight.

Calculations and experimental investigations of this problem without the restriction of minimizing the influence of the liquid are described in [29]. The principal propagation constant – the complex wave number $k = k_1 + ik_2$, by which the attenuation, phase velocity, and the amplitude distribution of the displacements in the liquid and solid [Eqs. (I.44), (I.45)] can then be computed – was calculated on the Ural computer for various parameters of the solid and liquid. The results are shown in Figs. 21-24. The curves are constructed for values of $r = c_t/c_L$ in the range 1.5-10. The set of values used for ρ_L/ρ, ν, and r ran the gamut for all practical situations. It is evident from Figs. 21a and 21b that the presence of a liquid at the boundary of the elastic half-space increases the surface wave velocity in the half-space, this increase being more pronounced the greater the ratio ρ_L/ρ. The dependence of the velocity increment on r and ν is also monotonic – as r and ν are increased, the increment decreases. It is important to note that the increase in surface-wave velocity due to the influence of the liquid is rather small; for average values of the parameters when $\rho_L/\rho = 0.5$, $\nu = 0.25$, and $r = 5$, the relative increase amounts to 0.0012, i.e., about 0.1%. It follows from Figs. 22a and 22b that the surface wave attenuation factor increases monotonically with increasing ρ_L/ρ and decreasing r and ν. The influence of the liquid on the damping of surface waves (by contrast with its influence on the velocity) is very pronounced; for the same typical values of the parameters ρ_L/ρ, r, and ν the coefficient of excess attenuation due to radiation into the liquid over a path λ_R amounts to 0.11, i.e., the wave is attenuated by a factor of 1/e over a path of approximately ten wavelengths.

The experimental investigations on the propagation of ultrasonic surface waves at an interface with a liquid were carried out on a pulsing apparatus consisting of a rectangular pulse generator with a sinusoidal carrier, an amplifier, and an indicator. The measurements were performed at frequencies of 1, 2, and 3 Mc with pulse durations of 10-50 μsec. Ultrasonic surface waves were excited by the "wedge" method and were propagated over the surfaces of rectangular aluminum and steel bars. The conditions for the propagation of surface waves at the interface between two half-spaces were simulated by immersing one end of the bar in a liquid path. Then the surface waves, in going from one face of the bar endwise to the other face, were brought in contact with the liquid for part of the way. By varying the depth of immersion, the attenuation and change in phase velocity of the surface wave due to the influence of the liquid could be determined. The attenuation was measured by the pulsing technique, the velocity by the pulse-phase method. The experimental and analytical data on the attenuation and velocity are shown in Table 4 for two bars (aluminum and steel) and three liquids (water, transformer oil, glycerin). As the table shows, the agreement between the experimental and analytical data is fair: within 15% for the velocity, 10% for the attenuation.

We turn now to a consideration of plane harmonic surface waves at the interface between a solid half-space and a plane liquid layer of thickness h, the second boundary of which is free. As before, we will be primarily concerned with waves that transform to Rayleigh waves in the solid as the density of the liquid tends to zero. A theoretical and experimental investigation of such waves is described in [29]. We will summarize its results. We introduce a system of coordinates with origin on the surface of the half-space, the x axis coinciding, as before, with the direction of wave propagation, and the z axis

TABLE 4

Contacting media	$\left(\frac{c-c_R}{c_R}\right)$, theor. %	$\left(\frac{c-c_R}{c_R}\right)$, % exp.			$k_2\lambda_R$ theor.	$(k_2\lambda_R)$ exp.		
		f= 1 Mc	f= 2 Mc	f= 3 Mc		f= 1 Mc	f= 2 Mc	f= 3 Mc
Steel – water	0.054	0.062	0.062	0.061	0.070	0.073	0.069	0.075
Steel – transformer oil	0.042	0.046	0.051	0.046	0.057	0.061	0.061	0.059
Steel – glycerin	0.190	0.13	0.14	0.13	0.124	0.11	0.12	0.11
Aluminum – water	0.480	0.49	0.52	0.50	0.189	0.20	0.19	0.20
Aluminum – transformer oil	0.350	0.38	0.35	0.35	0.157	0.17	0.16	0.16
Aluminum – glycerin	1.550	1.05	1.00	1.15	0.352	0.30	0.29	0.25

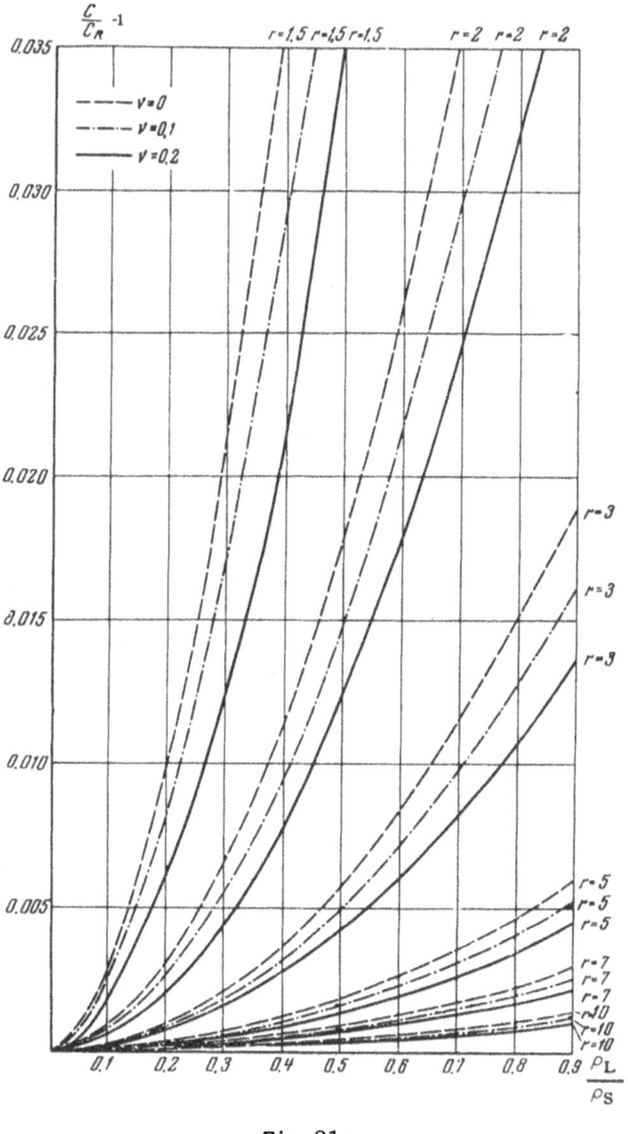

$\frac{C}{C_R}-1$
0.035
0.030
0.025
0.020
0.015
0.010
0.005
ν = 0
ν = 0.1
ν = 0.2
r=1.5 r=1.5 r=1.5
r=2 r=2 r=2
r=3
r=3
r=3
r=5
r=5
r=5
r=7
r=7
r=7
r=10
r=10
r=10
0.1
0.2
0.3
0.4
0.5
0.6
0.7
0.8
0.9
$\frac{\rho_L}{\rho_S}$

Fig. 21a

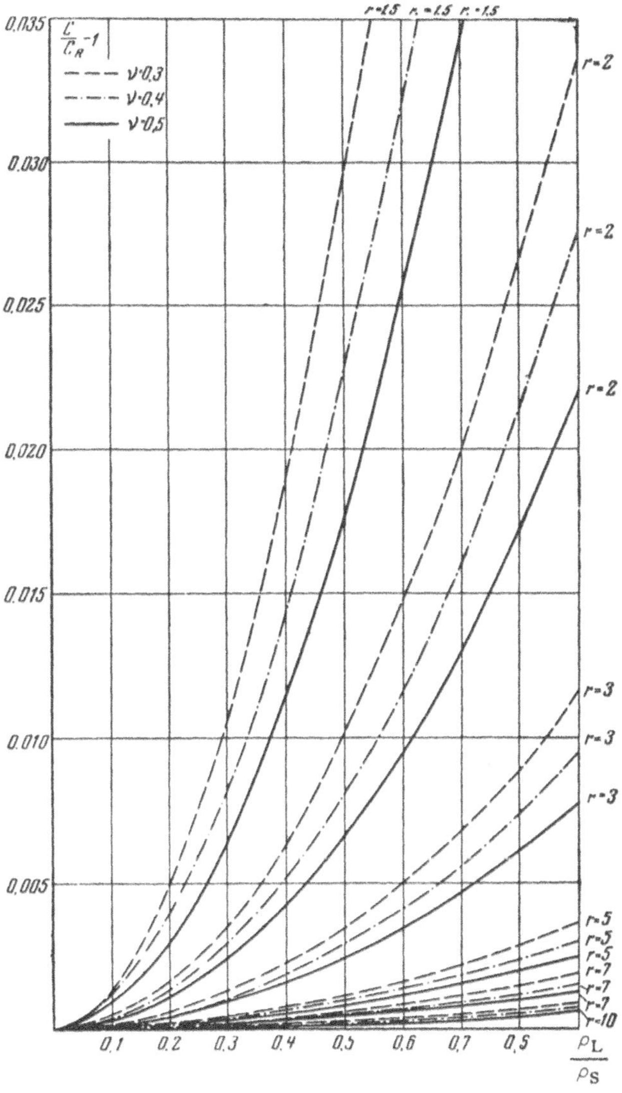

$\frac{C}{C_R}-1$
ν=0.3
ν=0.4
ν=0.5
0.035
0.030
0.025
0.020
0.015
0.010
0.005
0.1
0.2
0.3
0.4
0.5
0.6
0.7
0.8
$\frac{\rho_L}{\rho_S}$
r=1.5
r=1.5
r=1.5
r=2
r=2
r=2
r=3
r=3
r=3
r=5
r=5
r=5
r=7
r=7
r=7
r=10

Fig. 21b

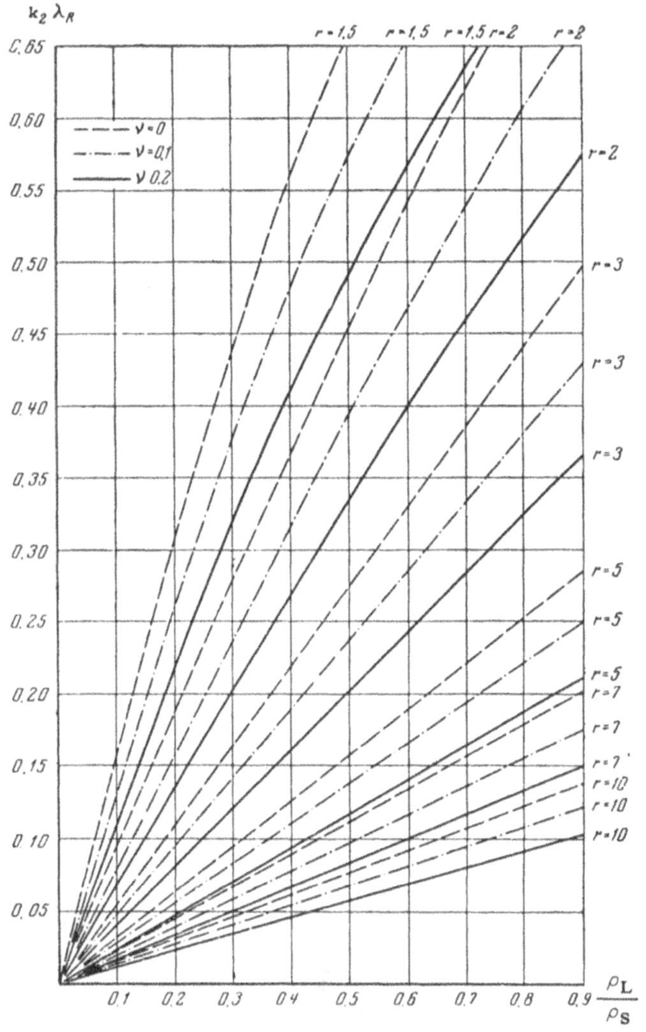

k2 λR
0.65
0.60
0.55
0.50
0.45
0.40
0.35
0.30
0.25
0.20
0.15
0.10
0.05
ν = 0
ν = 0.1
ν 0.2
r = 1.5
r = 1.5
r = 1.5
r = 2
r = 2
r = 2
r = 3
r = 3
r = 3
r = 5
r = 5
r = 5
r = 7
r = 7
r = 7
r = 10
r = 10
r = 10
0.1
0.2
0.3
0.4
0.5
0.6
0.7
0.8
0.9
ρL/ρS

Fig. 22a

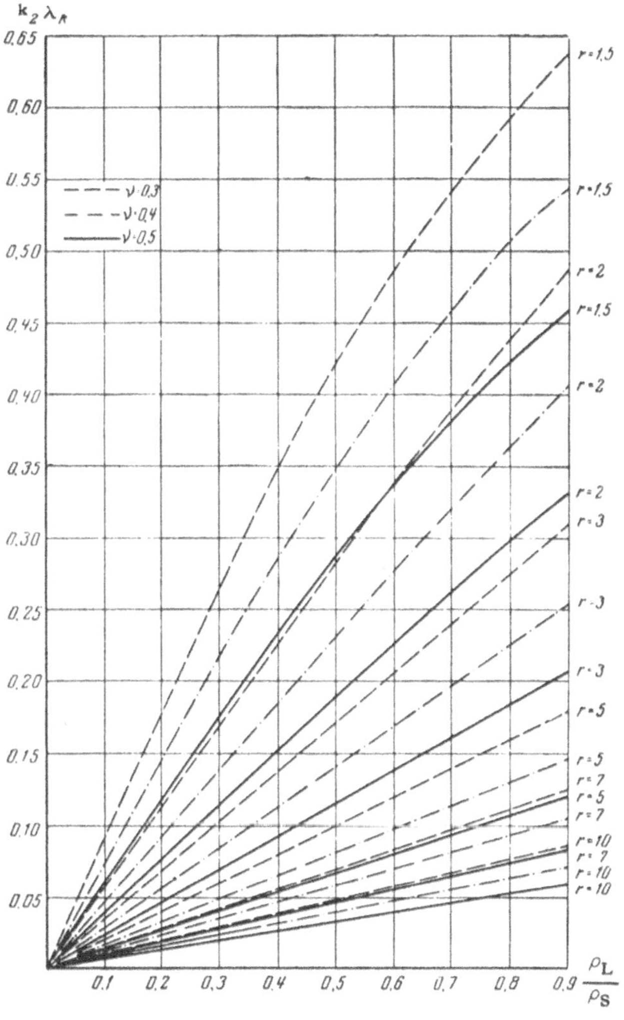
$k_2 \lambda_R$
0.65
0.60
0.55
0.50
0.45
0.40
0.35
0.30
0.25
0.20
0.15
0.10
0.05
ν=0.3
ν=0.4
ν=0.5
r=1.5
r=1.5
r=2
r=1.5
r=2
r=2
r=3
r=3
r=3
r=5
r=5
r=7
r=5
r=7
r=10
r=7
r=10
r=10
0.1
0.2
0.3
0.4
0.5
0.6
0.7
0.8
0.9
$\frac{\rho_L}{\rho_S}$

Fig. 22b

being directed into the depth of the half-space. Repeating the arguments of the first part of this section and performing some straightforward computations, we obtain the following expressions in place of Eqs. (I.44):

$$U_L = Ak\frac{qk_t^2}{\sqrt{k_L^2 - k^2}\,(k^2+s^2)} \cdot \frac{\sin\left[\sqrt{k_L^2 - k^2}\,(h+z)\right]}{\cos\sqrt{k_L^2 - k^2}\,h} \times$$
$$\times\, e^{i\left(kx-\omega t+\frac{\pi}{2}\right)}; \qquad \text{(I.47)}$$
$$W_L = Ak_t\frac{qk_t}{(k^2+s^2)^2} \cdot \frac{\cos\left[\sqrt{k_L^2 - k^2}\,(h+z)\right]}{\cos\sqrt{k_L^2 - k^2}\,h}\, e^{i(kx-\omega t)}.$$

The expressions for the displacements in the solid are described again by Eqs. (I.45). The characteristic equation for the given case has the form

$$(k^2+s^2)^2 - 4k^2qs = \frac{\rho_L q}{\rho\sqrt{k_L^2 - k^2}}\tan\sqrt{k_L^2 - k^2 h}. \qquad \text{(I.48)}$$

The results of a rigorous numerical solution of this equation are shown in Fig. 23. The ratio of the layer thickness to the wavelength in the liquid is plotted on the horizontal axis, the relative difference in velocities of the surface and Rayleigh waves on the vertical. One of the analytical curves (1) corresponds to a layer of transformer oil on steel, the other (2) to the case of a liquid and solid medium with "typical" parameters $\nu = 0.25$, $r = 3$, $\rho_L/\rho = 0.40$. Also shown in Fig. 23 is an experimental curve (1') recorded for a layer of transformer oil on steel. As seen in the graphs, with an increase in the layer thickness from zero to $h \approx (1/4)\lambda_L$, the velocity of the surface wave increases from c_R to c_t. In the vicinity of $h = (1/4)\lambda_L$, which corresponds to the maximum effect exerted by the liquid on the surface wave, there are no real values of the velocity, i.e., it is impossible for a propagating surface wave to exist (the curve suffers a discontinuity). To the right of the region of discontinuity the surface-wave phase velocity curve has two branches, blending smoothly into one another and corresponding to two surface modes. The lower branch of the curve tends asymptotically to the value $(c_L - c_R)/c_R$. At $\lambda/\lambda_L = 1/2\sqrt{1-k_R^2/k_L^2}$ the surface wave velocity becomes equal to c_R, i.e., the liquid "stops influencing" the surface wave in the solid. It can be shown that with a further increase in h/λ_L the upper branch of the curve behaves like the lower branch in the interval $h/\lambda_L \in 0 - 0.25$, then again the curve suffers a discontinuity, and so on. The set of curves obtained as h/λ_L varies from 0 to ∞ corresponds to the set of permissible normal modes of various orders

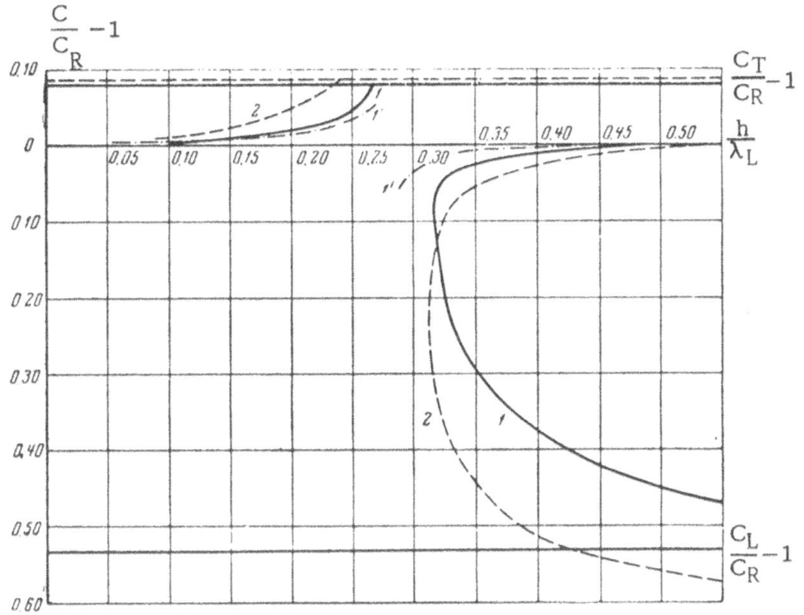

Fig. 23

that can exist in the layer. At $h/\lambda_L = n/2\sqrt{1-k_R^2/k_L^2}$ (n = 0, 1, 2,...) the curves intersect the horizontal axis ($c = c_R$).

For the experimental investigation of surface wave propagation at an interface between a solid and liquid layer (carried out on the same apparatus as the investigation of the solid and liquid half-spaces), a layer of liquid 0 to 0.7 mm thick was deposited on a section of the lateral surface of a steel bar, and the change of phase of the Rayleigh wave transmitted through this section was measured. The experiments were conducted at a frequency of 1 Mc with a pulse duration of 50 μsec. From the change of phase the variation in velocity was calculated (see curve 1' in Fig. 23). Curve 1' represents the average of thirty separate curves obtained in the measurements. It is obvious from its comparison with the corresponding theoretical curve that the qualitative behavior of both curves is the same, but the quantitative discrepancies are rather substantial. This is attributable to experimental errors, above all, clearly, the impracticability of creating a liquid layer of constant thickness (so that the oil would not run off the surface of the bar, the section to be covered with oil had to be bordered with a wire frame, thus creating pronounced menisci at the edges). It is important to note that the surface wave corresponding to the lower branch of the phase velocity curve (see Fig. 23) was not observed in

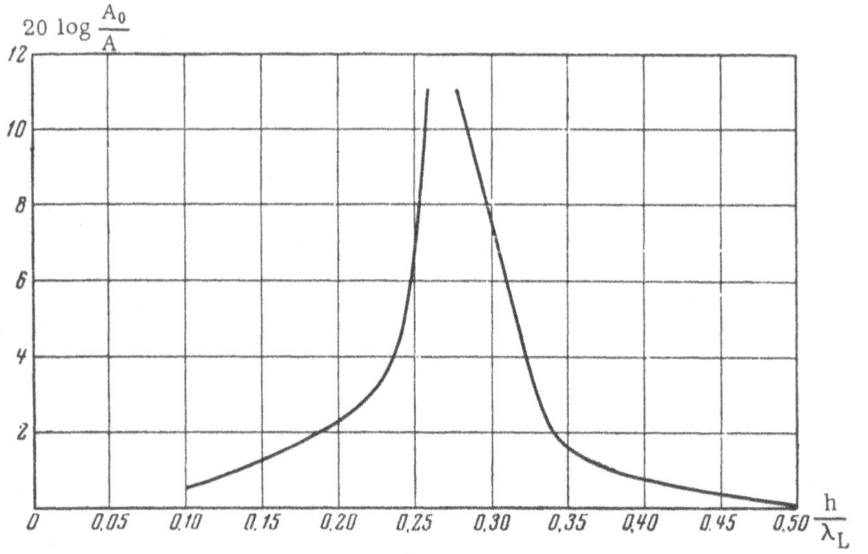

Fig. 24

our experiments. This can only be explained by the rather large departure of this wave from a Rayleigh wave (in its velocity and other characteristics), so that the Rayleigh wave generated by our emitter could only have been moderately transformed into the indicated surface mode on reaching the section covered with the oil layer.

If the liquid layer does not cover the entire surface of the solid, but only an isolated section of length l, a Rayleigh wave propagating over the free surface and passing through this section will be transformed into a surface wave at the boundary of the solid and liquid layer, as a result of which the energy of the Rayleigh wave is diminished (this explains the suppressant action of liquid films and surface contaminations on a Rayleigh wave). We investigated losses of this nature. As it turned out, for a layer long enough ($l \approx 5\text{-}10\lambda_R$) that a traveling surface wave has time to build up, the losses do not depend on the length of the layer or the pulse duration. This shows that the principal loss mechanism must lie in losses due to wave transformation and the formation of a normal mode in the liquid layer, rather than viscous losses. The dependence of the losses on the thickness of the layer is very significant. The attenuating effect of a film of transformer oil of length $l \approx 7\lambda_R$ on steel is illustrated in Fig. 24 for various thickness h (A_0 is the Rayleigh wave amplitude without the oil film, A is the same amplitude with film). The curve represents the average of thirty measurements, carried out at a frequency of 1 Mc and pulse duration of 50 μsec. As apparent from the curve, the maximum attenuation of the Rayleigh wave due to the liquid layer occurs at $h = \lambda_L/4$, i.e., in the same region where the surface-wave phase velocity curve passes through a discontinuity.

§ 7. Effect of Surface Defects on the Propagation of Rayleigh Waves

When defects of any sort (cracks, scratches, irregularities, etc.) are present on the surface over which a Rayleigh wave is propagating, the wave is scattered, as are longitudinal and transverse (shear) waves propagating into the depth of the medium. The initial Rayleigh wave is attenuated due to these scattering effects. The effect of surface defects on the propagation of Rayleigh waves is a very important problem with regard to the practical utilization of these waves, particularly in application to ultrasonic surface-flaw detection. This problem has been investigated in a number of papers.

The attenuation of Rayleigh waves during propagation over an uneven surface whose profile is described by an equation of the form $z = \zeta(x, y)$ is discussed theoretically in [30]. We present the fundamental results of this work. Let a solid medium occupy the half-space $z > \zeta$. We assume that the depth of the irregularities is small in comparison with λ_R and that the slopes of the surface are small with respect to the median plane. We solve the problem by the method of successive approximations, to and including the first approximation, assuming that in the zeroth approximation a plane harmonic Rayleigh wave propagates along the plane boundary in the positive direction. In the first approximation it is necessary to make allowance for the fact that the wave propagates along an uneven surface, as a result of which the boundary conditions of the problem call for zero stress on the surface $z = \zeta(x, y)$, instead of the plane $z = 0$. The stresses σ_{xz}, σ_{yz}, σ_{zz} will have nonzero values at points on the surface $z = \zeta(x, y)$ and can be expressed in terms of the stresses σ_{xx}^0, σ_{yy}^0, σ_{zz}^0, σ_{xz}^0 in the initial Rayleigh wave of the zeroth approximation, to wit:

$$
\begin{aligned}
\sigma_{xz} &= \left(\frac{\partial \sigma_{xz}^0}{\partial z}\right)_{z=0} \zeta - (\sigma_{xx}^0)_{z=0} \frac{\partial \zeta}{\partial x}; \\
\sigma_{yz} &= -(\sigma_{yy}^0)_{z=0} \frac{\partial \zeta}{\partial y}; \\
\sigma_{zz} &= -\left(\frac{\partial \sigma_{zz}^0}{\partial z}\right)_{z=0} \zeta.
\end{aligned}
\tag{I.49}
$$

We now "transfer" these stresses from the points of the uneven surface $z = \zeta(x, y)$ to the plane $z = 0$, assuming that the effect of the surface irregularities on the generation of Rayleigh waves is equivalent to the action of the indicated stresses on the plane $z = 0$. After this, the problem of waves scattered by surface irregularities reduces to the problem of vibration of a solid half-space with a plane boundary, on which the stresses are specified by (I.49).

We assume that the function $\zeta(x, y)$ may be represented by a double Fourier series:

$$\zeta = \sum_{m,n=-\infty}^{+\infty} \zeta_{mn} e^{i(mg_x x + ng_y y)}, \tag{I.50}$$

where $g_x = 2\pi/\Lambda_x$, $g_y = 2\pi/\Lambda_y$, and Λ_x and Λ_y are the periods of the surface irregularities in the x and y directions. Each pair of numbers m, n, i.e., each term of the series (I.50), corresponds to a pair of scattered waves (mn), one of which is longitudinal and is described by the scalar potential

$$\varphi_{mn} = A_{mn} e^{i(k_m x + k_n y + \sqrt{k_l^2 - k_m^2 - k_n^2}\, z - \omega t)}, \tag{I.51}$$

the other is transverse and is described by the vector potential

$$\Psi_{mn} = \mathbf{B}_{mn} e^{i(k_m x + k_n y + \sqrt{k_t^2 - k_m^2 - k_n^2}\, z - \omega t)}, \tag{I.52}$$

where $k_m = k_R + mg_x$, $k_n = ng_y$. For real $\sqrt{k_l^2 - k_m^2 - k_n^2}$ and $\sqrt{k_t^2 - k_m^2 - k_n^2}$ these waves travel in the direction away from the boundary, taking away a part of the energy of the fundamental Rayleigh wave. For imaginary radicals the corresponding waves will be surface waves propagating along the boundary with a velocity differing from the velocity c_R of the fundamental Rayleigh wave. The wave amplitude A_{mn} and B_{mn} may be determined by equating the stresses created by the set of waves (mn) to the stresses (I.49).

Every pair of waves (mn), in transporting the energy of the Rayleigh wave away from the boundary, makes its own contribution to the attenuation of the fundamental wave. We suppose that the attenuation of the Rayleigh wave due to scattering is sufficiently small that over some interval $l \gg \lambda_R$ in the direction of wave propagation, containing a large number of irregularities, the variation in amplitude of the fundamental wave is relatively small. In this case we may bring in the concept of a "partial attenuation factor" δ_{mn}, equal to the ratio of the energy withdrawn along the z axis per unit area of the wave (mn) to the energy flux density in the Rayleigh wave. The total attenuation factor δ for the Rayleigh wave (with respect to energy) is then equal to

$$\delta = \sum_{mn} \delta_{mn}. \tag{I.53}$$

The equation for δ_{mn} is quite cumbersome and we will not give it here. The results of calculating the attenuation factor for aluminum (curve 1), soil (curve 2), and steel (curve 3) are shown in Fig. 25 for the simplest type of surface irregularity, a sinusoidal corrugation, when $\zeta = \zeta_0 \cos gx$. The ordinates in Fig. 25 represent the dimensionless variable $\gamma = \delta/k_t^3\zeta_0^2$. When the irregularities have a spatial period $\Lambda = \lambda_R/(1 \pm c_R/c_l)$, each of the curves has very sharp maxima, near which the attenuation increases greatly (by a factor of e over a

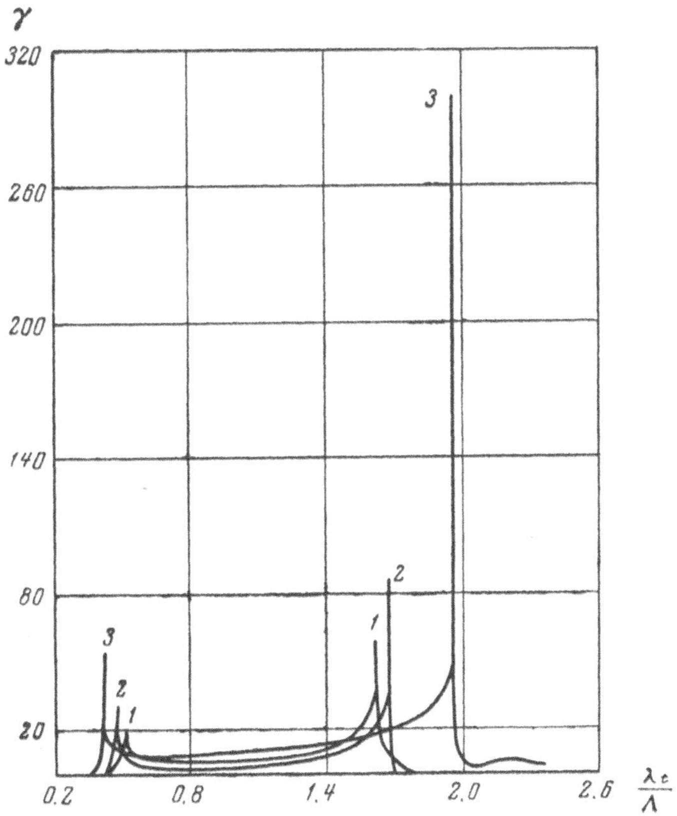

Fig. 25

path equal to 5-10λ_R). For these values of Λ the scattered longitudinal waves propagate in the same direction as the primary Rayleigh wave (left peaks) or in the opposite direction (right peaks). The steep increase in scattering at these values of Λ is caused by the efflux of energy from the boundary into the scattered transverse wave. For Λ greater than some Λ_{max} (for aluminum $\Lambda_{max} = 12.740\lambda_t$, for soil $10.26\lambda_t$, and for steel $4.895\lambda_t$), attenuation of the Rayleigh wave is generally absent.

The process of surface finishing of a metal (planing, milling, grinding, and polishing) and its effect on the attenuation of Rayleigh waves was investigated experimentally in [31]. It is shown that the attenuation depends very significantly on the method of finishing or machining, because with different finishing processes a different surface structure and surface layer is obtained; cold hardening, machining lines, different orientations on the part of structural elements in the surface layer of the metal, and surface films (in chemical polishing) all occur.

An experimental investigation of the effect of isolated surface defects on the propagation of plane Rayleigh waves was conducted in [32]. The investigation was carried out with two models of line surface defects: slots cut in the surface of wave propagation and semicylindrical grooves in the surface. Both the slot and the groove were oriented perpendicular to the direction of propagation of the Rayleigh wave. The majority of elongated surface defects (cracks, depressions, scratches in the surface) may be simulated by the two given defect models. In the investigation, the coefficient of reflection K_{re} of a Rayleigh wave from the model, the transmission coefficient K_{tr} across it, and the sum of the squares $K_{re}^2 + K_{tr}^2$, characterizing the transformation of the Rayleigh wave into volume waves scattered by the model, were measured. The measurements were performed according to the same scheme as in the experiments described in Section 5 on propagation along dihedral surfaces. The notches and grooves were cut into the flat surfaces of Dural bars.

The results of the measurements for the notch are shown in Fig. 26a, for the groove in Fig. 26b. The depth of the slot in Rayleigh wavelengths h/λ_R and ratio of the radius of the semicylindrical groove to the Rayleigh wavelength R/λ_R are plotted on the horizontal axes. It is evident from these figures that the sum $K_{re}^2 + K_{tr}^2$ is considerably less than unity for both the slot and the groove, testifying to the sizable transformation of energy that takes place from the Rayleigh wave into waves scattered by the slot or groove. As the depth of the slot and radius of the groove are increased, the coefficients of reflection of a Rayleigh wave from these defect models increase in an oscillating manner, while the transmission coefficients decrease in an oscillating manner (we note that the amplitudes of the oscillations are substantially greater

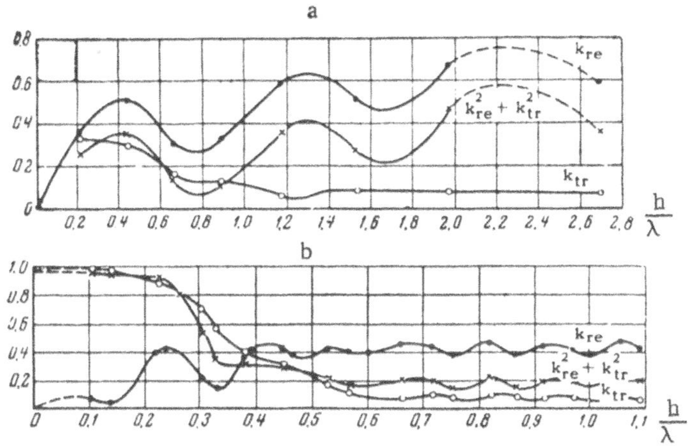

Fig. 26

than the error in measuring the reflection and transmission coefficients). With increasing h/λ_R the mean value of K_{re} for the slot tends to the value of the reflection coefficient for a Rayleigh wave at a wedge with apex angle $\theta = 90°$ (see Fig. 19). The transmission coefficient for a Rayleigh wave across the slot with $h/\lambda_R > 1.5$ no longer depends on the depth and is approximately equal to 0.08; consequently, a Rayleigh wave can pass over a crack of any depth, a fact which is important in ultrasonic flaw detection. For identical slot and groove depths, given the condition $h, R < 0.4\lambda_R$, the shielding and reflecting powers of the slot are greater than for the groove, hence a shallow surface crack is more readily disclosed by the flaw detector than a depression. The oscillations of K_{re} and K_{tr} for the slot and groove are attributable to the interference mechanisms involved in the formation of the reflected and transmitted Rayleigh waves. In the case of the slot the reflected wave is built up by reflection from the leading face (toward the emitter) of the slot and by scattering from the "bottom" of the slot of that "part" of the Rayleigh wave which "descended" there along the leading face (therefore the alternation of maxima and minima in K_{re} for the slot occurs at approximately every interval $\Delta h \simeq \lambda_R$). The transmitted wave is built up by direct transmission beneath the slot of the lower "part" of the incident Rayleigh wave not intercepted by the slot and by bending around the slot of the second (upper) part of the Rayleigh wave, which is transmitted across the slot, "descending" along one face and "ascending" along the other (for this reason the period of oscillation of K_{tr} is half that of K_{re}). In the case of the groove the reflected and transmitted waves are caused by the interference of reflections from the two rims of the groove – the leading and trailing edges. For a shallow groove ($R \ll \lambda_R$), the difference of these reflections is equal to $8\pi R/\lambda_R$, since the incident Rayleigh wave passes across the groove without altering its direction of propagation. We therefore have a reflection minimum at $R/\lambda_R \simeq 0.125$ and a miximum at $R/\lambda_R \simeq 0.250$. We note that a reflection mechanism similar to the one described is possible for shallow grooves of any configuration – for example, a rectangular groove.

For a groove with depth of the order λ_R or more, the incident Rayleigh wave and the wave reflected from the rear edge of the groove bend around the groove, propagating along its perimeter. In this case, the transmission and reflection of the waves become perfectly analogous to the transmission and reflection at the rounded edges of a rectangular wedge (see Section 5). Since surface waves of the Rayleigh type are strongly attenuated on a concave cylindrical surface with $R \sim \lambda_R$, the transmission coefficient for Rayleigh waves across a groove with $R/\lambda_R > 0.6$ is very small. We note that the behavior of the curves for the reflection and transmission coefficients for a slot and a groove, as implied by the described mechanisms for the formation of the reflected and transmitted Rayleigh waves, must be the same for any elastic medium.

An experimental investigation of the scattering of ultrasonic Rayleigh waves by another two types of surface-defect models is described in [33]; one is a hemispherical recess of varying diameter and a cylindrical channel of varying diameters and depths, cut perpendicularly from the surface of Rayleigh wave propagation. These models, which comprise a natural complement to the first two, may be thought of as localized surface defects of the same type as pits – vertical cracks running from the surface, etc. The models, together with the models of elongated defects, more or less characterize the entire gamut of surface defects. The measurements were performed in the pulsed mode with a pulse duration of 10 μsec and carrier frequency of 2.75 Mc. Defect models of various dimensions were made on the well finished surfaces of rectangular Dural sheets with dimensions 450 × 300 × 7 mm. The emission and reception of Rayleigh waves were realized by the wedge method. The emitting wedge was placed at a distance of 225 mm from the defect model and propagated into it a directional beam of Rayleigh waves. The receiving wedge was placed successively at points on a circle of radius 50 mm inscribed about the model, each measurement of the scattered wave amplitude being referred at once to a corresponding measurement of the incident wave amplitude at some point between the emitter and defect model. The amplitude of the incident Rayleigh wave right at the model was determined from the latter measurement by conversion. The receiving wedge was acoustically coupled with the surface of the Dural sheet only along a circle 3 mm in diameter, thus enabling the amplitude of the sheet surface vibrations to be measured within a

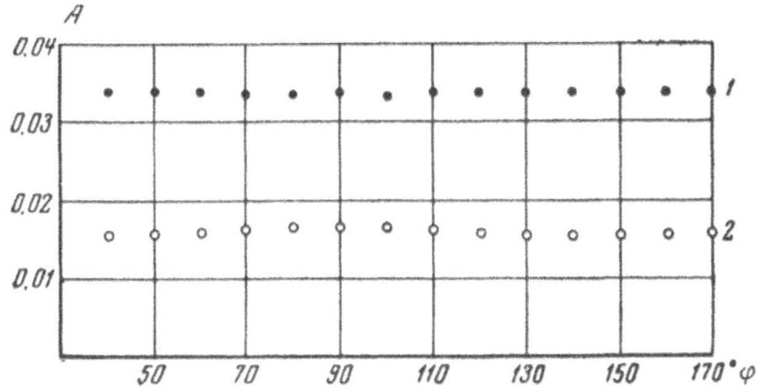

Fig. 27

small region (locally). Acoustic coupling was established by a film of castor oil.

The results of the measurements are shown in Figs. 27, 28a, and 28b. Figure 27 illustrates the dependence of the amplitudes A of Rayleigh waves scattered by a cylindrical channel (1) of diameter $d = 1.8$ mm and depth $h = 3$ mm and by a hemispherical recess (2) of diameter $d = 1.6$ mm on the angle ϕ measured clockwise from the direction of the wave vector of the Rayleigh wave incident on the model. For other model dimensions the dependence of the scattered wave amplitudes on the angle ϕ was exactly the same, hence the corresponding data are not presented. The dependences of the amplitudes of Rayleigh waves scattered by the cylindrical channel (1) and hemispherical recess (2) on the ratio d/λ_R are shown in Fig. 28a (the amplitudes were measured for $40° < \phi < 170°$, where they do not depend on ϕ). The dimensionless amplitude A represents the ratio of the surface displacement amplitude in the Rayleigh wave scattered by the model at a distance $r = 50$ mm from the model to the surface displacement amplitude in the incident Rayleigh wave at the model, all multiplied by the attenuation factor. The attenuation factor compensates for the decrease in amplitude of the scattered wave on the path from the model to the point of reception (50 mm) due to absorption and scattering in the Dural. The dependence of the amplitude A on the ratio of the depth of the cylindrical channel to the wavelength is shown in Fig. 28b.

The first thing we note from Fig. 27 is that in the angular interval $40° < \phi < 170°$ the scattering by the cylindrical channel and hemispherical recess is the same in all directions (measurements were not made for $\phi \sim 180°$, $\phi < 40°$, because the reflected pulse was masked by the incident pulse). The same is true, of course, for negative values of ϕ as well. This result is quite unexpected and is evidently related to the specific nature of the Rayleigh wave, in that ordinarily for the scattering of waves by obstacles in liquid or

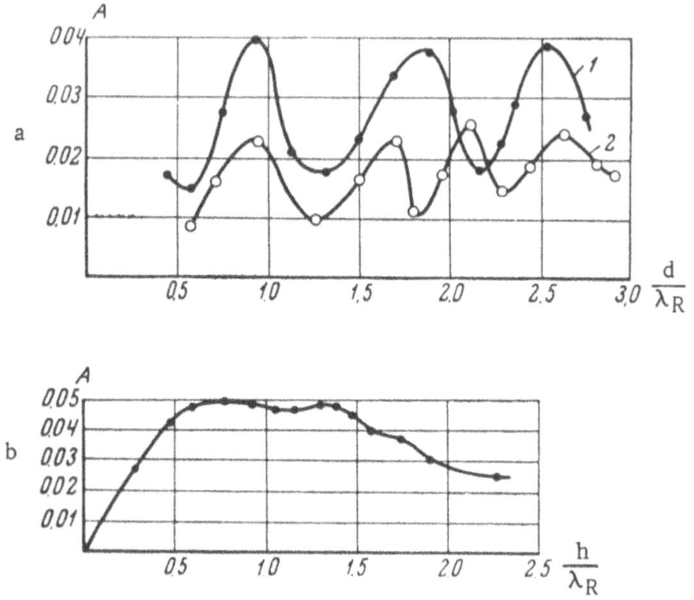

Fig. 28

solid, where the dimensions of the obstacles are comparable with the wavelength, the scattering intensity is strongly dependent on the angle ϕ.

It follows from Fig. 28a that the dependence of the amplitude of Rayleigh waves scattered by the models on the ratio d/λ_R is not monotonic, but oscillates considerably, where the period of these oscillations for a cylindrical channel and hemispherical recess of diameters $d < 1.7\lambda_R$ amounts to about 0.80; for a hemispherical recess with diameter $d > 1.7\lambda_R$ the period is approximately equal to 0.45. It may be assumed, by analogy with the scattering of longitudinal waves by a cylindrical cavity in a solid, that the scattering maxima in our case correspond to resonances of the cylindrical and hemispherical cavities (channel and recess). The cylindrical and hemispherical cavities of small diameter ($d < 1.7\lambda_R$) resonate at identical values of d, which indicates, clearly, resonances of the perimeter πd in these cases. For $d > 1.7\lambda_R$, the hemispherical recess discloses new resonances associated with the specific nature of its geometry by comparison with the cylindrical channel. Comparing curves 1 and 2 in Fig. 28a, we note that the cylindrical cavity scatters Rayleigh waves far more intensely than the hemispherical cavity. For scattering by a cylindrical channel the amplitude of the scattered Rayleigh wave at first increases with increasing depth of the channel, then decreases in a moderately oscillating fashion (see Fig. 28b).

This behavior on the part of the dependence $A(h/\lambda_R)$ fosters the assumption that in the given instance the scattered wave is formed as the result of radiation of two in-phase scattering centers, one of which occupies a region near the upper edge of the channel, the other near the "bottom." If the channel depth is greater than the thickness of the layer in which the Rayleigh mode is localized ($h > 2\lambda_R$), the "bottom" source "cuts out," causing the amplitude of the scattered wave to drop. The waviness of the curve is clearly related to interaction between the scattering centers, which is brought about through a standing Rayleigh wave occurring at the wall of the channel between its upper edge and the bottom. This assumption is supported by the length of the spatial period exhibited by the waviness, which is approximately equal to $\lambda_R/2$.

The scattering of Rayleigh waves by models of localized surface defects is conveniently evaluated by the introduction of an effective scatterer dimension. We will interpret the effective dimension D for both models as the width of that part of the beam of Rayleigh waves impinging on the model (the width is measured right at the model) in which is concentrated an energy equal to the total energy of the scattered Rayleigh waves. Consequently, D is understood to mean the dimension (in the direction perpendicular to the wave vector of the incident wave) of some ideal reflector of Rayleigh waves. With D so defined and assuming, in addition, that the Rayleigh wave amplitude is the same for all angles ϕ, we have the following expression for D:

$$D = 2\pi r A^2,$$

where r is the distance from the center of the model at which the amplitude of the scattered Rayleigh wave (in the experiments described herein r = 50 mm). Computing the effective scattering dimensions of the models according to this formula, we find, for example, that the maximum value of D = 0.85 mm corresponds to a cylindrical channel 0.5 mm in diameter and 0.83 mm in depth, the minimum value of 0.025 mm corresponding to a hemispherical recess of diameter d = 0.7 mm. The parameter D is a universal intrinsic characteristic of a defect, determined solely by its shape and the ratio of the linear dimensions to the incident Rayleigh wavelength. Knowing the value of D and the amplitude B of the incident wave at the defect, it is possible to calculate the amplitude b of the scattered wave at any distance R from the defect, according to the formula

$$b = \sqrt{D/2\pi R}\, B$$

(the formula is applicable, of course, with the stipulation that the scattering amplitude is independent of the angle ϕ).

Chapter II

PHYSICAL PROPERTIES OF ULTRASONIC LAMB WAVES

§ 1. The Concept of Lamb Waves

Lamb waves refer to elastic perturbations propagating in a solid plate (or layer) with free boundaries, for which displacements occur both in the direction of wave propagation and perpendicularly to the plane of the plate. Lamb waves represent one of the types of normal or plate modes in an elastic waveguide, in this case a plate with free boundaries. For this reason, Lamb waves are sometimes simply called normal modes in a plate. But this definition is rather loose, insofar as another type of normal mode can exist in a plate with free boundaries, namely transverse normal modes, wherein the motion is perpendicular to the direction of propagation and parallel to the boundaries of the plate.

We will consider, after the pattern of [3], a plane harmonic Lamb wave propagating in a plate of thickness 2d in the positive x direction (Fig. 29). For the region occupied by the plate we introduce the scalar potential ϕ and vector $\boldsymbol{\psi}$ of the displacements, which are related to the particle displacements by the relation (I.1). As in the case of a plane Rayleigh wave (see Section 1, Chapter I), the vector potential has a nonzero magnitude only in the direction of the y axis, hence we refer only to that component, or simply ψ. The potentials ϕ and ψ, which describe longitudinal and transverse waves, respectively, must satisfy the wave equations (I.2). The connection between these potentials and the displacement components U and W along the x and z axes, as well as the components of the stress tensor is given by Eqs. (I.3) and (I.4).

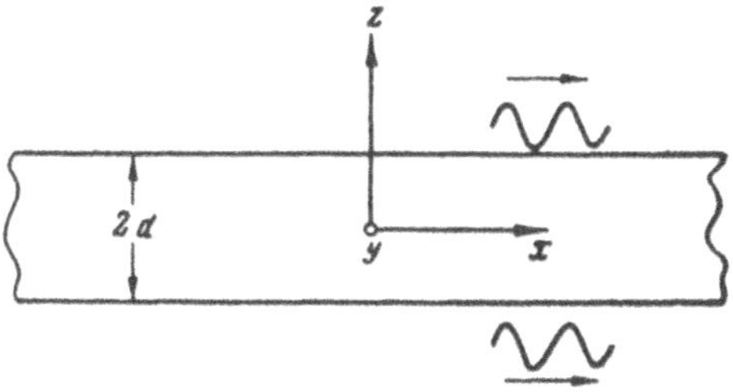

Fig. 29

We represent φ and ψ in the following form:

$$\begin{aligned}\varphi &= A_s \operatorname{ch} qze^{ikx} + B_a \operatorname{sh} qze^{ikx},\\ \psi &= D_s \operatorname{sh} sze^{ikx} + C_a \operatorname{ch} sze^{ikx},\end{aligned} \tag{II.1}$$

where A_s, B_a, C_a, D_s are arbitrary constants, k is the Lamb wave number, $q=\sqrt{k^2-k_l^2}$; $s=\sqrt{k^2-k_t^2}$; the factor $e^{-i\omega t}$ is dropped for brevity. It is not too difficult to see that these expressions satisfy the wave equations (I.2). Moreover, they must cause the stresses σ_{xz} and σ_{zz} on the planes $z = \pm d$ to go to zero. Substituting Eqs. (II.1) into Eq. (I.4) and equating the indicated stresses to zero, we obtain the following system of linear homogeneous equations for the amplitudes A_s, B_a, C_a, D_s:

$$\begin{aligned}
&(k^2+s^2)\operatorname{ch} qdA_s + (k^2+s^2)\operatorname{sh} qdB_a + 2iks\operatorname{sh} sdC_a + 2iks\operatorname{ch} sd\,D_s = 0;\\
&(k^2+s^2)\operatorname{ch} qdA_s - (k^2+s^2)\operatorname{sh} qdB_a - 2iks\operatorname{sh} sdC_a + 2iks\operatorname{ch} sdD_s = 0;\\
&2ikq\operatorname{sh} qdA_s + 2ikq\operatorname{ch} qdB_a - (k^2+s^2)\operatorname{ch} sdC_a - (k^2+s^2)\operatorname{sh} sdD_s = 0;\\
&-2ikq\operatorname{sh} qdA_s + 2ikq\operatorname{ch} qdB_a - (k^2+s^2)\operatorname{ch} sdC_a + (k^2+s^2)\operatorname{sh} sdD_s = 0.
\end{aligned}$$

It is readily seen that the above system is satisfied if the following two subsystems are satisfied:

$$\begin{aligned}&(k^2+s^2)\operatorname{ch} qdA_s + 2iks\operatorname{ch} sdD_s = 0,\\ &2ikq\operatorname{sh} qdA_s - (k^2+s^2)\operatorname{sh} sdD_s = 0;\end{aligned} \tag{II.2}$$

$$\begin{aligned}&(k^2+s^2)\operatorname{sh} qdB_a + 2iks\operatorname{sh} sdC_a = 0,\\ &2ikq\operatorname{ch} qdB_a - (k^2+s^2)\operatorname{ch} sdC_a = 0.\end{aligned} \tag{II.3}$$

The subsystems have nontrivial solutions only when their determinants are equal to zero. This leads to two characteristic equations, determining the eigenvalues of the wave number k:

$$(k^2+s^2)^2\operatorname{ch} qd\operatorname{sh} sd - 4k^2qs\operatorname{sh} qd\operatorname{ch} sd = 0; \tag{II.4}$$

$$(k^2+s^2)^2\operatorname{sh} qd\operatorname{ch} sd - 4k^2qs\operatorname{ch} qd\operatorname{sh} sd = 0. \tag{II.5}$$

With these equations we obtain from the subsystem (II.2) an expression for D_s in terms of A_s, and from the subsystem (II.3) an expression for C_a in terms of B_a. Inserting them into (II.1), we arrive at the following relations for the desired potentials:

$$\varphi = A_s \operatorname{ch} q_s z e^{ik_s x} + B_a \operatorname{sh} q_a z e^{ik_a x};$$
$$\psi = \frac{2ik_s q_s \operatorname{sh} q_s d}{(k_s^2 + s_s^2) \operatorname{sh} s_s d} A_s \operatorname{sh} s_s z e^{ik_s x} + \frac{2ik_a q_a \operatorname{ch} q_a d}{(k_a^2 + s_a^2) \operatorname{ch} s_a d} B_a \operatorname{ch} s_a z e^{ik_a x}. \quad \text{(II.6)}$$

Here, k_s are the values of k satisfying Eq. (II.4), k_a are the values satisfying Eq. (II.5); $q_{s,a} = \sqrt{k_{s,a}^2 - k_l^2}$; $s_{s,a} = \sqrt{k_{s,a}^2 - k_t^2}$. The displacement components U and W may be calculated from Eqs. (II.6) with the help of the relations (I.3):

$$U = U_s + U_a; \quad \text{(II.7)}$$
$$W = W_s + W_a,$$

where

$$U_s = Ak_s \left(\frac{\operatorname{ch} q_s z}{\operatorname{sh} q_s d} - \frac{2q_s s_s}{k_s^2 + s_s^2} \cdot \frac{\operatorname{ch} s_s z}{\operatorname{sh} s_s d} \right) e^{i\left(k_s x - \omega t - \frac{\pi}{2}\right)},$$
$$W_s = -Aq_s \left(\frac{\operatorname{sh} q_s z}{\operatorname{sh} q_s d} - \frac{2k_s^2}{k_s^2 + s_s^2} \cdot \frac{\operatorname{sh} s_s z}{\operatorname{sh} s_s d} \right) e^{i(k_s x - \omega t)}; \quad \text{(II.8)}$$

$$U_a = Bk_a \left(\frac{\operatorname{sh} q_a z}{\operatorname{ch} q_a d} - \frac{2q_a s_a}{k_a^2 + s_a^2} \cdot \frac{\operatorname{sh} s_a z}{\operatorname{ch} s_a d} \right) e^{i\left(k_a x - \omega t - \frac{\pi}{2}\right)},$$
$$W_a = -Bq_a \left(\frac{\operatorname{ch} q_a z}{\operatorname{ch} q_a d} - \frac{2k_a^2}{k_a^2 + s_a^2} \cdot \frac{\operatorname{ch} s_a z}{\operatorname{ch} s_a d} \right) e^{i(k_a x - \omega t)}; \quad \text{(II.9)}$$

here A and B are new arbitrary constants.

The expressions (II.6)-(II.9) and Eqs. (II.4), (II.5) describe two groups of waves, each of which satisfies the wave equations of motion and boundary conditions, i.e., both can propagate in the plate independently of one another. Analyzing the expressions (II.8) and (II.9), we see right away that the first group of waves, indicated by the subscript s, describes waves in which the motion occurs symmetrically with respect to the plane $z = 0$ (i.e., the displacement U has the same signs, the displacement W opposite signs in the upper and lower halves of the plate). The second group, indicated by the subscript a, describes waves in which the motion is antisymmetrical with respect to $z = 0$ (i.e., the displacement U has opposite signs, the displacement W the same signs in the upper and lower halves of the plate). The waves of the first group

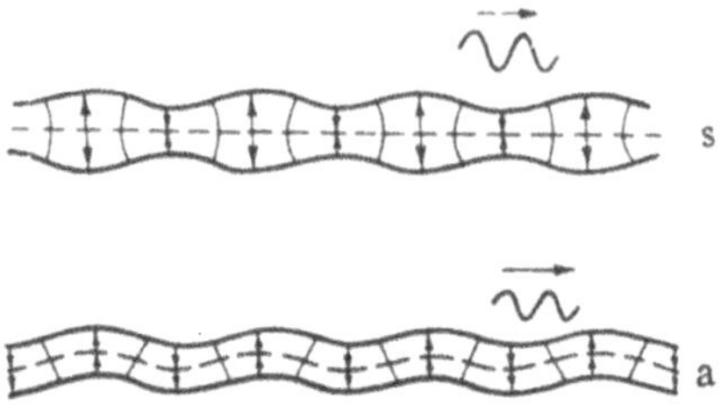

Fig. 30

are called symmetrical Lamb waves, those of the second group are called antisymmetrical. The deformation of the plate in the z direction during propagation of symmetrical (s) and antisymmetrical (a) waves therein is illustrated in Fig. 30.

§ 2. The Number of Lamb Waves and Critical Frequencies

In a plate of thickness 2d at a frequency ω there can exist a finite number of symmetrical and antisymmetrical Lamb waves, differing from one another by their phase and group velocities and distribution of the displacements and stresses throughout the thickness of the plate. The number of symmetrical waves is determined by the number of real roots of Eq. (II.4), the number of antisymmetrical waves by the real roots of Eq. (II.5). Every root defines a wave number $k_{s,a}$ or phase velocity $c_{s,a}$ of the corresponding wave. It can be shown that, in addition to the finite number of real roots $k_{s,a}$, both of the indicated equations have for any ω and d an infinite set of purely imaginary roots, corresponding to in-phase motions of the plate which decay or grow exponentially along the x axis. Since we are concerned with propagating waves, we will only deal with the real roots of the characteristic equations (II.4) and (II.5).

For $\omega d \to 0$, Eqs. (II.4) and (II.5) clearly have only one root each. The root of Eq. (II.4) corresponds to the so-called zeroth symmetrical normal mode, which we will designate s_0, while the root of Eq. (II.5) represents the zeroth antisymmetrical mode a_0. As ωd increases, the roots k_{s_0} and k_{a_0} vary in magnitude, and for definite ratios between ω and d new roots appear, corresponding to the first, second, and higher symmetrical ($s_1, s_2, \ldots, s_n$) and antisymmetrical ($a_1, a_2, \ldots, a_n$) Lamb waves. The values of ω and d at which new roots appear are called the "critical" thicknesses and frequencies. The relations between the critical thicknesses and transverse and longitidunal

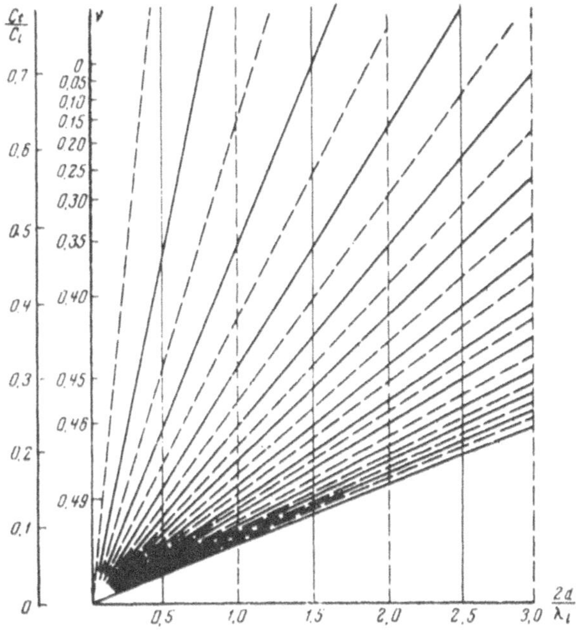

Fig. 31

wavelengths are as follows:

$$\left.\begin{aligned} 2d &= \frac{\lambda_l}{2},\ \frac{3\lambda_l}{2},\ \frac{5\lambda_l}{2},\ldots \\ 2d &= \lambda_t,\ 2\lambda_t,\ 3\lambda_t,\ldots \end{aligned}\right\}\quad \text{for symmetrical modes;} \qquad \text{(II.10)}$$

$$\left.\begin{aligned} 2d &= \lambda_l,\ 2\lambda_l,\ 3\lambda_l,\ldots \\ 2d &= \frac{\lambda_t}{2},\ \frac{3\lambda_t}{2},\ \frac{5\lambda_t}{2},\ldots \end{aligned}\right\}\quad \text{for antisymmetrical modes.} \qquad \text{(II.11)}$$

At critical frequencies the wave numbers $k_{s,a} \to 0$ (the phase velocities $c_{s,a} \to \infty$), and the new symmetrical or antisymmetrical mode represent a standing longitudinal [upper set of values in Eqs. (II.10) and (II.11)] or transverse [lower set of values in (II.10) and (II.11)] wave in the plate.

The total number of symmetrical modes N_s that are possible in a plate of given thickness 2d at the frequency ω is equal to

$$N_s = 1 + \left[\frac{2d}{\lambda_t}\right] + \left[\frac{2d}{\lambda_l} + \frac{1}{2}\right], \qquad \text{(II.12)}$$

the total number of antisymmetrical modes

$$N_a = 1 + \left[\frac{2d}{\lambda_l}\right] + \left[\frac{2d}{\lambda_t} + \frac{1}{2}\right]. \tag{II.13}$$

The brackets in this case indicate the nearest integer part of the number that they enclose. A nomogram, borrowed here from [34], is shown in Fig. 31 for the determination of the number of Lamb waves in a plate. The ratio of the plate thickness to the longitudinal wavelength is plotted on the horizontal axis, the velocity ratio $c_t/c_l = \xi$ (uniform scale) and Poisson ratio ν (nonuniform scale) are shown on the vertical. The vertical lines correspond to the critical frequencies of waves occurring as longitudinal modes, the sloping lines correspond to the critical frequencies of waves occurring as transverse modes (in both cases the solid lines represent symmetrical modes, the dashed lines antisymmetrical modes). In order to determine the number of propagating waves corresponding to a given value of $2d/\lambda_l$, one counts the number of lines corresponding to a particular type of wave to the left of the point corresponding to a given ratio $2d/\lambda_l$ and Poisson ratio ν for the given material. It is apparent from the graph that as the Poisson ratio is increased, the number of waves appearing as transverse modes increases over the number of waves appearing as longitudinal modes.

§ 3. Phase and Group Velocities of Lamb Waves

The phase velocity c is the fundamental characteristic of a Lamb wave. Once it is known, we can determine the wave number and calculate the stresses and displacements in the Lamb wave at any point of the plate. Furthermore, it is possible from the relation $c^{gr} = (c - \lambda)(dc/d\lambda)$ to calculate the group velocity of the wave, c^{gr}. The phase velocity is found by numerical solution of the characteristic equations (II.4) and (II.5), which are suitably rewritten for the purpose of these calculations in the following dimensionless form:

for symmetrical modes:

$$\frac{\tan\sqrt{1-\zeta^2}\,\overline{d}}{\tan\sqrt{\xi^2-\zeta^2}\,\overline{d}} = -\frac{4\zeta^2\sqrt{1-\zeta^2}\sqrt{\xi^2-\zeta^2}}{(2\zeta^2-1)^2},$$

for antisymmetrical modes:

$$\frac{\tan\sqrt{1-\zeta^2}\,\overline{d}}{\tan\sqrt{\xi^2-\zeta^2}\,\overline{d}} = -\frac{(2\zeta^2-1)^2}{4\zeta^2\sqrt{1-\zeta^2}\sqrt{\xi^2-\zeta^2}};$$

where

$$\bar{d} = k_t d; \quad \zeta^2 = \frac{c_t^2}{c^2}; \quad \xi^2 = \frac{c_t^2}{c_l^2}.$$

Many authors have performed calculations of the phase and group velocities and their dependence on the plate thickness and frequency (dispersion curves). At the Acoustics Institute of the Academy of Sciences of the USSR, the Ural computer was used to run a detailed calculation of the dispersion curves for the first (lowest) twenty five Lamb modes [35]. We have reproduced these curves for the phase velocities of the modes s_0, s_1, s_2, a_0, a_1, a_2 in Fig. 32a and for the group velocities of the same modes in Fig. 32b. The curves correspond to a Poisson ratio $\nu = 0.34$.

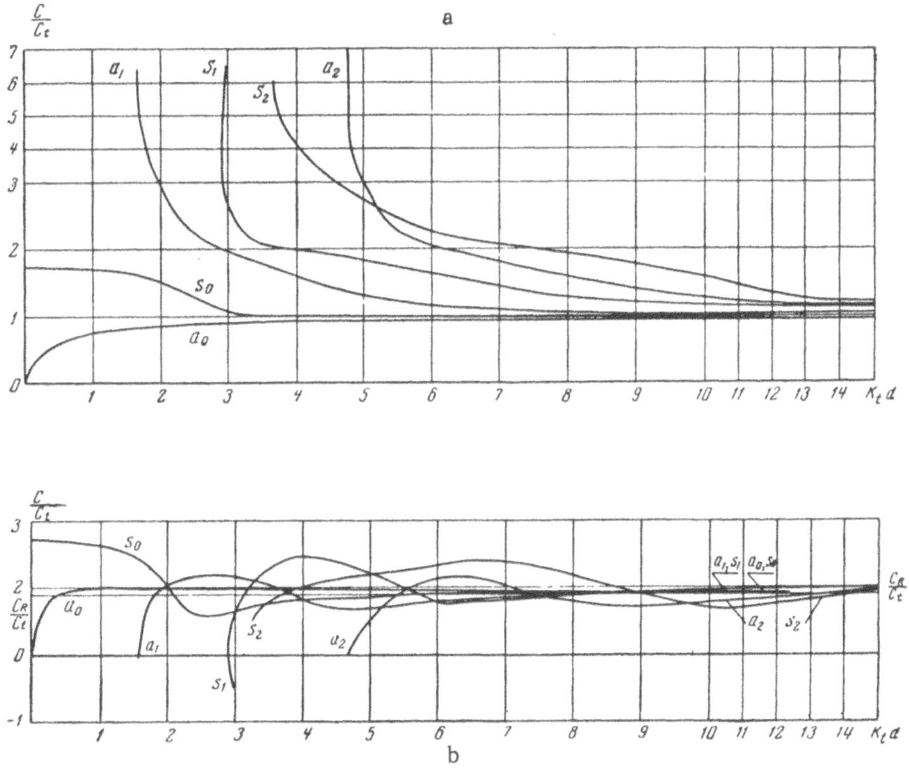

Fig. 32

As seen in Figs. 32a and 32b, powerful velocity dispersion is typical of waves of every order. As $\bar{d}$ varies from zero to infinity, the phase velocities of the s_0 and a_0 waves vary respectively from

$$\sqrt{E/\rho(1-\nu^2)} \quad \text{and} \quad \sqrt[4]{E/3\rho(1-\nu^2)}\sqrt{\omega d}$$

(where E is Young's modulus) to c_R. The group velocities of these waves vary accordingly from

$$\sqrt{E/\rho(1-\nu^2)} \quad \text{and} \quad 2\sqrt[4]{E/3\rho(1-\nu^2)}\sqrt{\omega d}$$

to c_R. The phase velocities of the nonzero modes vary from infinity (at the instant the wave sets in) to $c \to c_t$ for $\bar{d} \to \infty$. The group velocities of the nonzero normal modes vary from zero and even negative values (sometimes) to values near c_t, for $\bar{d} \to \infty$. A negative group velocity means that the phase and group velocities of the given wave are in the opposite direction. It is evident from the graphs that every wave of order higher than zero has phase-velocity maximum. It is shown in [34] that for certain modes there can be several maxima in the group velocity, the value of the largest maximum increasing with the order number of the wave and approaching c_l. Moreover, with increasing wave order, near the largest maximum of the symmetrical modes there appear intervals of $\bar{d}$ in which the group velocity remains almost constant. For example, in the case of the mode s_{10}, the relative interval of $\bar{d}$ in which the group velocity does not vary by more than 3% turns out to be about 10%. In the indicated intervals the phase velocity also changes very little. The intervals of quasiconstant group velocity values are followed by typical sharp drops in c^{gr}.

At the critical values of $k_t d$, which we designate $\bar{d}_k$, the following asymptotic expressions are valid for the phase velocities of the new waves [36]:

1) symmetrical and antisymmetrical modes, generated as longitudinal waves:

$$c_s = \frac{c_l}{\sqrt{\bar{d}-\bar{d}_k}}\sqrt{\frac{\bar{d}_k}{2}+4\xi^2\cot\bar{d}_k}\,, \tag{II.14}$$

$$c_a = \frac{c_l}{\sqrt{\bar{d} - \bar{d}_k}} \sqrt{\frac{\bar{d}_k}{2} - 4\xi^2 \tan d_k}; \tag{II.15}$$

2) symmetrical and antisymmetrical modes, generated as transverse waves:

$$c_s = \frac{c_t}{\sqrt{\bar{d} - \bar{d}_k}} \sqrt{\frac{\bar{d}_k}{2} - 4\xi \tan \xi \bar{d}_k}, \tag{II.16}$$

$$c_a = \frac{c_t}{\sqrt{\bar{d} - \bar{d}_k}} \sqrt{\frac{\bar{d}_k}{2} + 4\xi \cot \xi \bar{d}_k}. \tag{II.17}$$

It is difficult to obtain asymptotic expressions for the phase velocities of the symmetrical and antisymmetrical modes of order n and for a very thick plate $k_t d \to \infty$. They have the form [36]

$$c_s = c_t \left[1 + \frac{1}{2} \left(\frac{n\pi}{\bar{d}} \right)^2 \right]; \tag{II.18}$$

$$c_a = c_t \left[1 + \frac{1}{8} \frac{(2n-1)^2 \pi^2}{\bar{d}^2} \right], \tag{II.19}$$

where n is the order number of the wave.

§ 4. The Motion in Lamb Waves

Knowing the dispersion curves for the phase and group velocities of Lamb waves, we are in a position to discuss the motion in waves of various order numbers. As a characteristic of the motion it is sufficient to know the displacement along the x and z axes at different points of the plate. The displacements can be used to calculate the stresses in the wave, making use of the following relations:

$$\begin{aligned}
\sigma_{xx} &= \lambda \left(\frac{\partial U}{\partial x} + \frac{\partial W}{\partial z} \right) + 2\mu \frac{\partial U}{\partial x}; \\
\sigma_{zx} &= \mu \left(\frac{\partial U}{\partial z} + \frac{\partial W}{\partial x} \right); \\
\sigma_{zz} &= \lambda \left(\frac{\partial U}{\partial x} + \frac{\partial W}{\partial z} \right) + 2\mu \frac{\partial W}{\partial z}.
\end{aligned} \tag{II.20}$$

We begin our discussion with the modes s_0 and a_0, which differ qualitatively from all the others in their velocity dispersion curves and in that they exist for any frequencies and plate thicknesses. For small plate thicknesses, when $k_{s,a_0} d \to 0$ these modes represent a longitudinal (s_0) and flexural (a_0) wave in a thin plate. Their displacements in this case are described by the following expressions, derived as the result of limiting transitions from Eqs. (II.8), (II.9):

$$U_{s_0} = Ak_t \frac{k_s^2 - s_s^2}{k_s^2 + s_s^2} \frac{1}{k_t d} \sin(k_s x - \omega t),$$
$$W_{s_0} = Ak_t \frac{q_s (k_s^2 - s_s^2)}{k_t (k_s^2 + s_s^2)} \frac{z}{d} \cos(k_s x - \omega t); \tag{II.21}$$

$$U_{a_0} = Bk_t \frac{k_t z}{2} \sin(k_a x - \omega t),$$
$$W_{a_0} = Bk_t \frac{k_t}{2k_a} \cos(k_a x - \omega t). \tag{II.22}$$

It is apparent from the equations that displacements along the x axis prevail in the longitudinal wave, the displacement amplitude being the same at all points of the plate. The displacement in the transverse direction due to the Poisson effect is less than the longitudinal displacement by a factor of approximately $1/k_t d$. It is a maximum on the surface and equal to zero in the median plane of the plate. In the flexural wave, on the other hand, the transverse displacement W prevails, its amplitude being the same at all points of the plate. The longitudinal displacement is equal to zero in the median plane and is a maximum on the surface of the plate; on the average it is less than the transverse displacement by a factor of $1/k_a d$. More detailed information may be found on the longitudinal and flexural modes in thin plates in the monograph [4].

As the thickness of the plate is increased, the properties of the waves s_0 and a_0 change; they become more and more "like" one another. For $k_t d \gg 1$ their phase velocities tend to the Rayleigh wave phase velocity c_R (see Fig. 32a), the displacements become localized near the free boundaries of the plate, and their distributions with depth tend to the depthwise distribution in a Rayleigh wave. The dependence of the amplitudes of the normal and tangential displacement in the modes s_0 and a_0 on the relative depth $(d-z)/\lambda_R$ for a half-plate has been calculated by us for the case $\nu = 0.34$, $d/\lambda_R = 1$ ($k_t d \simeq 6$) and is shown in Fig. 33. The curves are constructed after the example of Fig. 2; the displacement amplitudes are referred to the amplitudes of the normal

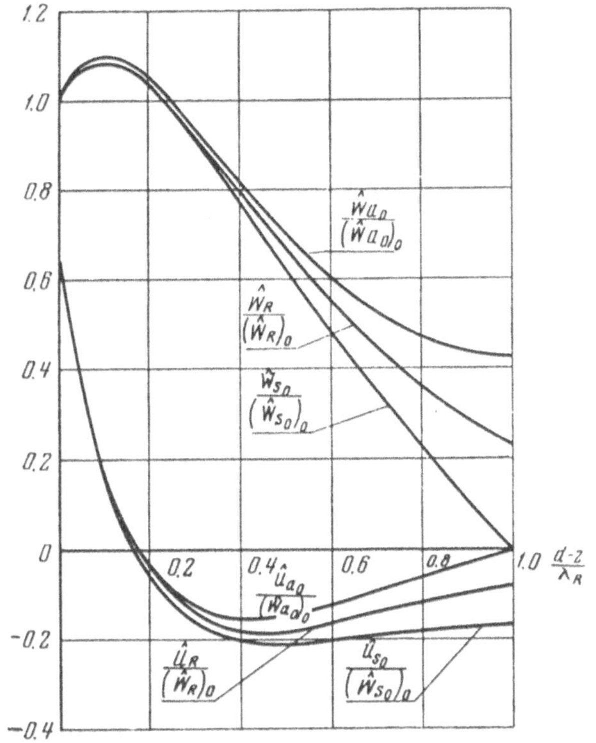

Fig. 33

displacements in the corresponding waves at the surface ($z = d$). Also plotted on the same graph is the dependence of the displacement amplitudes on the depth in a Rayleigh wave (curves with index R) for a medium with $\nu = 0.34$. As the graph shows, the largest discrepancies in the depthwise distributions of the displacements in the symmetrical, antisymmetrical, and Rayleigh modes are observed in the middle of the plate ($z \simeq 0$). Near each of the free surfaces of the plate all of the distributions are very similar. As $k_t d \to \infty$, each of the modes s_0 and a_0 go over to two Rayleigh waves propagating in phase along both free surfaces of the plate. This problem will be analyzed further in Section 6 of the present chapter.

We look now at the motion in waves of order higher than zero, which, as already mentioned, only appear for certain critical values of d_{cr}. For subcritical thicknesses and frequencies the phase velocities and wave numbers of these waves are purely imaginary. This means that there is no wave propagation, only in-phase motion of the particles all over the plate, this motion decaying exponentially in the direction of the x axis. At the critical values of $k_t d$, when an even or odd number of longitudinal or transverse half-waves fit into the thickness of the plate [see (II.10), (II.11)], the generated Lamb wave represents a purely longitudinal or purely transverse standing wave formed by two waves of corresponding polarizations propagating with equal amplitudes in the positive and negative directions of the z axis. The expression for the displacements in these cases are derived from Eqs. (II.8), (II.9) for $k_{s,a} \to 0$, taking into account the relations (II.10), (II.11), and have the following form:

Symmetrical and antisymmetrical modes, generated as longitudinal waves:

$$U_s = 0\left(\sqrt{\bar{d} - \bar{d}_k}\right), \quad W_s = Ak_l \sin k_l z \sin(kx - \omega t); \tag{II.23}$$

$$U_a = 0\left(\sqrt{\bar{d} - \bar{d}_k}\right), \quad W_a = Bk_l \cos k_l z \sin(kx - \omega t); \tag{II.24}$$

Symmetrical and antisymmetrical modes, generated as transverse waves:

$$U_s = \frac{Ak_l}{\sqrt{\bar{d}_k/2 - 4\xi \tan \xi \bar{d}_k}} \cos k_t z \cos(kx - \omega t), \quad W_s = 0\left(\sqrt{\bar{d} - \bar{d}_k}\right); \tag{II.25}$$

$$U_a = \frac{Bk_b}{\sqrt{\bar{d}_k/2 + 4\xi \cot \xi \bar{d}_k}} \sin k_t z \cos(kx - \omega t), \quad W_a = 0\left(\sqrt{\bar{d} - \bar{d}_k}\right). \tag{II.26}$$

For larger-than-critical values of $k_t d$, the Lamb wave numbers take on nonzero values. This may be interpreted as a rotation of the directions of propagation of the two longitudinal or transverse waves forming the standing wave in the critical region, from the z axis toward the positive x axis. Then, due to reflection from the plate boundaries, waves of a different polarization occur, and the Lamb wave is "composed" of four components (Fig. 34): two longitudinal waves (described by the potential φ) and two transverse waves (described by the potential ψ), which are "fitted" to one another so that the projections of all the wave numbers on the x axis are the same, while the stresses set up by the four waves on the bounding planes $z = \pm d$ are equal to zero. In this case, the distribution of the displacements throughout the thickness of the plate becomes complex and must be calculated according to the general equations (II.8), (II.9). The trajectories of the particles become ellipses, whose eccentricity depends on the type and order number of the wave, the depth, and the Poisson ratio of the plate material.

For certain $k_t d$ values, the wave vectors of the "longitudinal" components turn out to be directed exactly along the x axis. The wave vectors of the "transverse" components form a certain angle with the x axis at this time (always $k_l < k_t$). With a further increase in $k_t d$, the longitudinal components are transformed into inhomogeneous waves, which decay exponentially with distance from the boundaries $z = \pm d$ (as a result of which the projections of their wave numbers on the x axis "acquire the ability" to keep growing), and for the transverse components the wave vectors come nearer and nearer to the x axis. The displacements created by the longitudinal components become localized in surface layers of thickness $\lambda_t - 2\lambda_t$. Throughout the remaining thickness of the plate, the displacements are created solely by the transverse components. For $k_t d \to \infty$, the following simple expressions are obtained from (II.8), (II.9) for the components of these displacements (in the plate thickness) along the z axis:

$$W_s = A' k_t \sin\left(n\pi \frac{z}{d}\right) \cos(k_s x - \omega t),$$
$$W_a = B' k_t \cos\left(\frac{2n-1}{2} \cdot \frac{\pi z}{d}\right) \cos(k_a x - \omega t); \quad \text{(II.27)}$$

here A', B' are arbitrary constants and n is the wave number. The x components of the displacements U_s and U_a for $k_t d \to \infty$ are less than the z components by

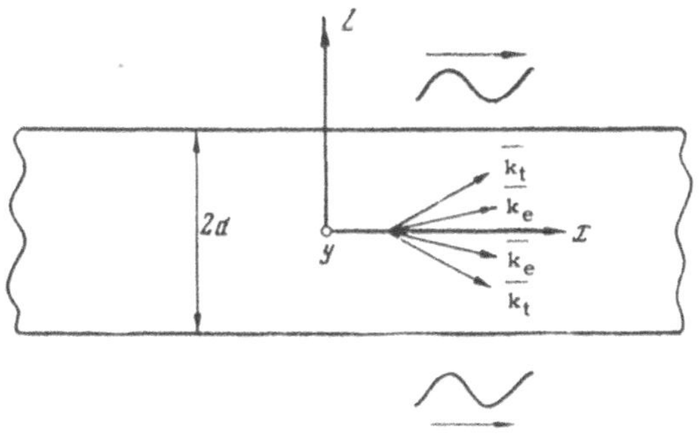

Fig. 34

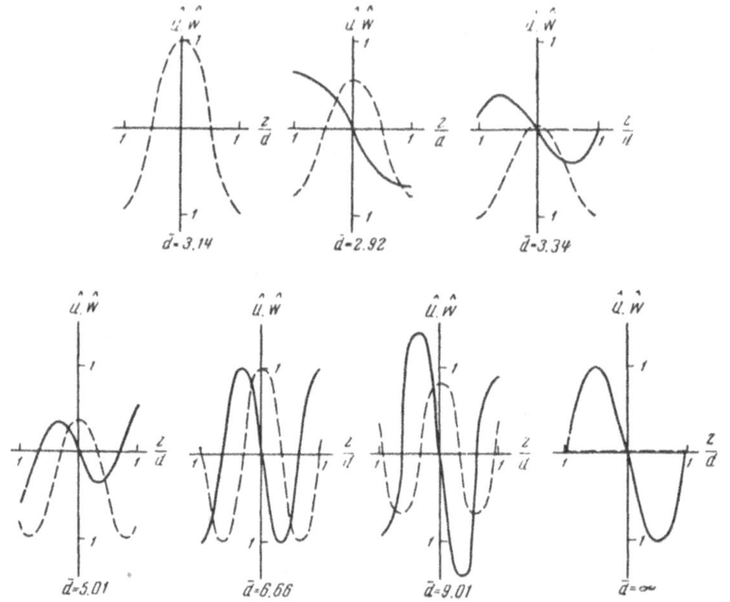

Fig. 35

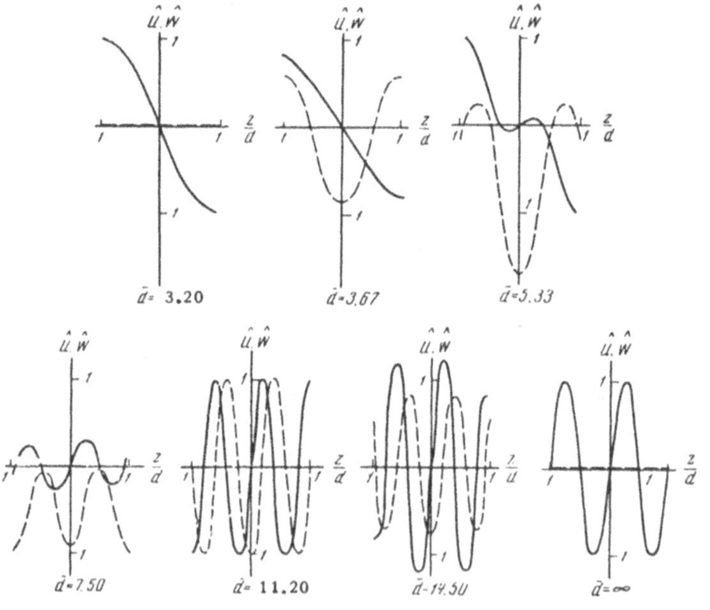

Fig. 36

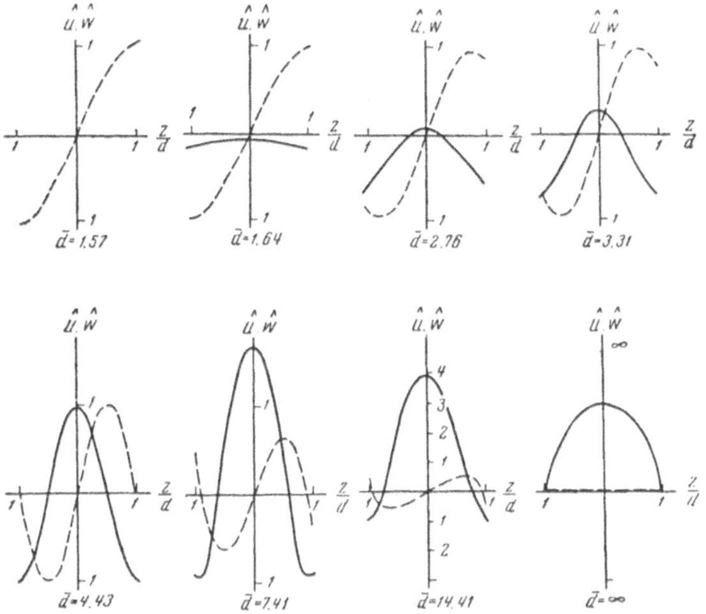
$\hat{u}, \hat{w}$
$\frac{z}{d}$
$\bar{d}$ = 1.57
$\bar{d}$ = 1.64
$\bar{d}$ = 2.76
$\bar{d}$ = 3.31
$\bar{d}$ = 4.43
$\bar{d}$ = 7.41
$\bar{d}$ = 14.41
$\bar{d}$ = ∞

Fig. 37

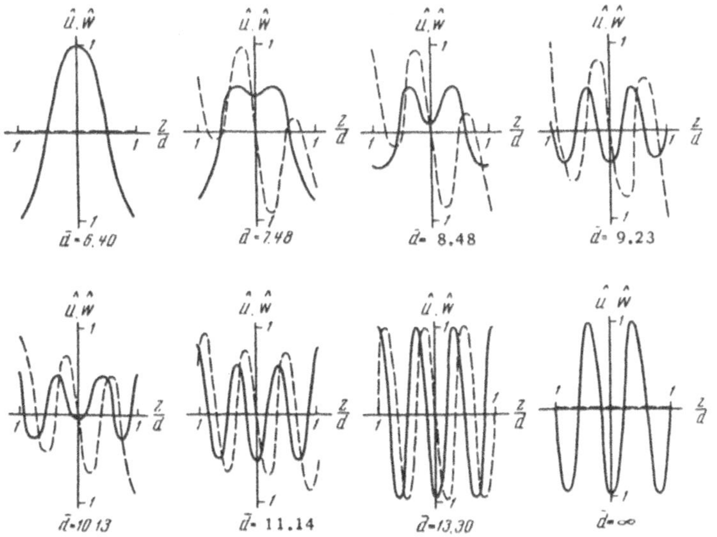
$\hat{u}, \hat{w}$
$\frac{z}{d}$
$\bar{d}$ = 6.40
$\bar{d}$ = 7.48
$\bar{d}$ = 8.48
$\bar{d}$ = 9.23
$\bar{d}$ = 10.13
$\bar{d}$ = 11.14
$\bar{d}$ = 13.30
$\bar{d}$ = ∞

Fig. 38

a factor of $1/k_t d$, hence they may be neglected. This means that for $k_t d \to \infty$, the displacements in all Lamb waves in the plate (except for the surface layers) are directed along the z axis and are distributed sinusoidally with spatial periods $2d/n$ (for symmetrical modes) and $2d/(n-{}^1/_2)$ (for antisymmetrical modes).

Our calculated distributions of the amplitudes of the tangential displacement $\hat{U}$ (dashed curves) and normal displacement $\hat{W}$ (solid curves) with depth in the first four modes s_1, s_2, a_1, a_3 are shown in Figs. 35-38, respectively. Two of these modes are symmetrical, two are antisymmetrical, where s_1, a_1 are produced as transverse waves, s_2, a_3 as longitudinal waves. We have skipped over the mode a_2, which like a_1, is created in the form of a transverse wave. The distributions have been constructed for each of the modes with seven or eight values of the relative plate thickness $\bar{d}$, beginning with the critical values corresponding to the onset of the mode and extending to $\bar{d} \to \infty$. The values of $\bar{d}$ are indicated in each graph. The distribution curves are drawn in dimensionless form; the depth z, measured from the median plane of the plate and plotted on the horizontal axes of all the graphs, is referred to the half-thickness of the plate d; the displacement amplitudes $\hat{U}$ and $\hat{W}$ along the x and z axes, which are plotted on the vertical axes, are referred to an average amplitude of the surface displacement, equal to the square root of the sum of the squares of the surface displacement amplitudes along the x and z axes. The curves have been calculated according to (II.8), (II.9) for values of the Poisson ratio $\nu = 0.34$.

It is apparent from Figs. 33-38 that for each mode the distributions of the displacement in the cross section of the plate represent rather complicated oscillating functions. In the light of the above discussion, the distributions have a simple form only at the instant a new wave appears and for $\bar{d} \to \infty$. In these situations the displacements amplitudes are distributed in a sine or cosine wave, and there is only one component of the displacement. At the instant a new wave appears this will be the z component if the wave is longitudinal, the x component if the wave is transverse. For $\bar{d} \to \infty$, all the waves have a diaplacement only along the z axis, the amplitude of this displacement tending on the surface to zero by comparison with the amplitude in the thickness of the plate. This means that for sufficiently large plate thicknesses the motion in each Lamb-wave mode, except s_0 and a_0, becomes localized in the thickness and does not "emerge" to the surface. For the modes s_0 and a_0, on the other hand, as already noted, a very characteristic skin effect occurs.

As the plate thickness is increased above the critical values, the number of oscillations in the cross-sectional displacement distribution functions increases, reaching a maximum at thickness defined by the condition

$k_{s,a}^2 + s_{s,a}^2 = 0$, after which the number of oscillations diminishes and the distribution functions become simpler, approaching their limit as $\overline{d} \to \infty$, With an increase in the order of the mode, the number of oscillations in the distribution function increases, and the distributions themselves become more complex.

§ 5. Methods for the Excitation and Reception of Lamb Waves

Two conceptually distinct methods are available for the excitation of Lamb waves: the creation of perturbations on the surface of a plate and the creation of perturbations inside a plate or at the end face. In practice, as a rule, the first possibility is used, as it is more convenient and simpler in its realization. We will therefore discuss only the first category of excitation methods. There are four well-known basic methods of excitation in this category, which are entirely analogous to the corresponding methods of Rayleigh wave excitation (see Section 2 of Chapter I and Fig. 5):

1) excitation by shearing perturbations, distributed uniformly over the surface of the plate in a strip of width 2a [37] and created by a Y-cut quartz plate or titanate plate of suitable polarization;

2) excitation by analogous normal perturbations, created by a quartz or titanate plate acoustically coupled to the surface of the test plate (so-called contact version of the given method of excitation) or by an ultrasonic beam at normal incidence on the plate in a liquid (immersion version, see [38]);

3) excitation by a set of normal perturbations, distributed periodically over the surface of the plate with a spatial period equal to the excited Lamb wavelength; this method, called the comb-structure method, was proposed for the excitation of ultrasonic Rayleigh waves, but it can be used with success for Lamb waves as well;

4) excitation of normal perturbations, distributed according to a sinusoidal traveling-wave law on the surface of the plate in some region of width 2b; in the contact version, this method is generally known as the "wedge method" (see [39, 40]). In the immersion version [41], the method is realized by the transmission of an ultrasonic beam at oblique incidence onto the plate from the liquid surrounding the latter.

The first and second methods of excitation are nonresonance methods; all possible normal modes for a given frequency are excited in approximately equal measure in the plate. This is usually not suitable for practical applications (for instance, in flaw detection, a whole series of signals are reflected from a defect, making it difficult to localize the defect). The comb-structure and wedge methods are of the resonance type; by varying the distance between the projections of the comb-shaped plate or the wedge angle, it is possible to excite any of the possible modes on a selective basis. In the wedge method, the "tuning" from mode to mode is realized very simply; it suffices merely to incorporate into the emitter design a provision for changing the angle θ. For this reason, the wedge method is widely used in ultrasonic flaw detection and other practical applications of Lamb waves. The comb-structure method is suitable for excitation of Lamb waves in plate materials with a low velocity of sound (such as plastic), since the application of the wedge method in this case is enormously complicated by the problem of finding a suitable material for the wedge, which of course must be chosen such that the longitudinal wave velocity in it is less than the Lamb-wave velocity in the plate. We note that all four of the enumerated methods (along with their singular characteristics) are reversible and may be used for the detection of Lamb waves.

The first method of excitation has been considered in detail in [37]; the second and third methods have been investigated theoretically in [36]. As already pointed out, all of the methods are analogous to their counterparts in the excitation of Rayleigh waves (in particular, the dependence of the amplitudes of the excited Lamb waves on the width of the excitation region is the same as for Rayleigh waves, except that k_R must now be replaced in the analogous equations by the wave numbers k_{s_n} or k_{a_n}, which are the same as for the equivalent Rayleigh waves, and allowance must be made for distortion of the pulse shape in the different methods, etc.). We will therefore not go into detail concerning all of these methods of exciting Lamb waves, but will confine our discussion solely to the one most widely used – the wedge method. This method has been investigated theoretically and experimentally in [42] and we give here its essential content.

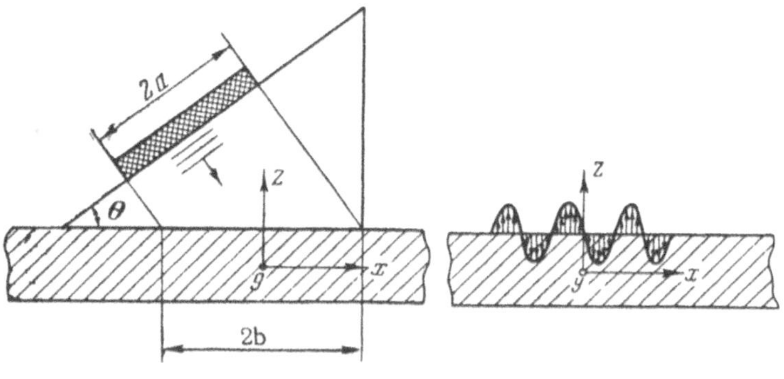

Fig. 39

For the theoretical discussion, we will consider the material of the plate to be homogenous and perfectly elastic, the dimension of the emitter along the y axis to be infinite, and its effect on the plate to be equivalent to the effect of a system of normal stresses applied to the upper free surface of the plate in the region $|x| \leq a/\cos\theta = b$, defined by the geometric boundaries of a beam of harmonic longitudinal waves of frequency ω propagating in the wedge (Fig. 39). The normal stresses will be assumed equal to the stresses arising when a plane wave is incident at an angle θ to the boundary between a solid half-space consisting of the wedge material and a solid layer (the plate), beyond which there is vacuum. Sliding acoustic contact is realized at the boundary, i.e., the normal stresses and displacements are equal in the half-space and layer at the boundary between them, and the tangential stresses there equal to zero. Sliding contact is realized in the majority of practical situations, because the emitter wedge and plate are normally coupled acoustically by means of a thin liquid film. The immersion version of the wedge method also meets these conditions, insofar as tangential stresses are not present at the boundary. The replacement of the prismatic emitter by a set of normal stresses, strictly speaking, is admissible provided the specific wave impedance of the emitter material is small in comparison with the specific wave impedance of the plate material. This is approximately true in the majority of practical situations (i.e., a plastic wedge on a metal plate. We note also that, given the aforementioned idealizations, we are neglecting the divergence of the beam of longitudinal waves in the wedge and the displacement of the reflected beam at the wedge-plate interface.

With the given assumptions, the problem reduces to an investigation of the vibrations of an infinite, plane solid layer, the lower surface of which ($z = -d$) is free, the upper ($z = d$), specified by the stresses

$$\sigma_{zz} = \begin{cases} \sigma_0 |\Phi| e^{i(k_0 x - \omega t)} & \text{for} \quad |x| \leqslant b, \\ 0 & \text{for} \quad |x| > b, \end{cases} \qquad \text{(II.28)}$$
$$\sigma_{xz} = 0 \qquad \text{for} \quad |x| < \infty;$$

here $k_0 = k_W \sin\theta$; k_W is the wave number for longitudinal waves in the wedge; σ_0 is the normal stress amplitude in a longitudinal wave in the wedge on an area of surface perpendicular to the direction of propagation (numerically this amplitude is equal to the amplitude of the stress created by a piezoelectric plate situated on the sloping face of the wedge); $|\Phi|$ is the modulus of the ratio of the stress amplitude σ_{zz} to σ_0 at $z = d$.

The analytical procedure used here is analogous to that used in the investigation of Rayleigh wave excitation. We will therefore only discuss it briefly. For the layer, we introduce the potentials ϕ and ψ for longitidinal and transverse waves, satisfying the wave equations (I.2). The displacement components U and W along the x and z axes are expressed in terms of ϕ and ψ according to Eqs. (I.3).

We represent ϕ and ψ by Fourier integrals, separating each of the potentials into symmetrical and antisymmetrical parts, such that the symmetrical parts of the potentials yield symmetrical motion relative to the plane $z = 0$ in the layer, the antisymmetrical parts yielding antisymmetrical motion:

$$\begin{aligned}\varphi &= \varphi_s + \varphi_a = \int_{-\infty}^{\infty} A(k)\,\mathrm{ch}\, qz e^{ikx} dk + \int_{-\infty}^{\infty} B(k)\,\mathrm{sh}\, qz e^{ikx} dk;\\ \psi &= \psi_s + \psi_a = \int_{-\infty}^{\infty} D(k)\,\mathrm{sh}\, sz e^{ikx} dk + \int_{-\infty}^{\infty} C(k)\,\mathrm{ch}\, sz e^{ikx} dk.\end{aligned} \tag{II.29}$$

Here A(k), B(k), C(k) are certain unknown functions; $q = \sqrt{k^2 - k_l^2}$; $s = k^2 - k_t^2$; the factor $e^{-i\omega t}$ is dropped for brevity. The functions A(k), B(k), C(k), D(k) are determined by the boundary conditions (II.28), writing the expressions for the stresses also in the form of Fourier integrals and expressing the stresses σ_{zz} and σ_{xz} in terms of ϕ and ψ. Then we can calculate the displacements at any point of the layer according to Eqs. (I.3):

$$\begin{aligned}U &= \frac{i\sigma_0 |\Phi|}{2\pi\mu} \int_{-\infty}^{\infty} k \left\{ \frac{(k^2 + s^2)\,\mathrm{sh}\, sd\,\mathrm{ch}\, qz - 2qs\,\mathrm{sh}\, qd\,\mathrm{ch}\, sz}{\Delta_s} + \right.\\ &\left. + \frac{(k^2 + s^2)\,\mathrm{ch}\, sd\,\mathrm{sh}\, qz - 2qs\,\mathrm{ch}\, qd\,\mathrm{sh}\, sz}{\Delta_a} \right\} \frac{\sin (k_0 - k)\, b}{k_0 - k} e^{ikx} dk,\\ W &= \frac{\sigma_0 |\Phi|}{2\pi\mu} \int_{-\infty}^{\infty} q \left\{ \frac{(k^2 + s^2)\,\mathrm{sh}\, sd\,\mathrm{sh}\, qz - 2k^2\,\mathrm{sh}\, qd\,\mathrm{sh}\, sz}{\Delta_s} + \right.\\ &\left. + \frac{(k^2 + s^2)\,\mathrm{ch}\, sd\,\mathrm{ch}\, qz - 2k^2\,\mathrm{ch}\, qd\,\mathrm{ch}\, sz}{\Delta_a} \right\} \frac{\sin (k_0 - k)\, b}{k_0 - k} e^{ikx} dk;\end{aligned} \tag{II.30}$$

where

$$\Delta_s = (k^2 + s^2)^2\,\mathrm{ch}\, qd\,\mathrm{sh}\, sd - 4k^2 qs\,\mathrm{ch}\, sd\,\mathrm{sh}\, qd;$$
$$\Delta_a = (k^2 + s^2)^2\,\mathrm{ch}\, sd\,\mathrm{sh}\, qd - 4k^2 qs\,\mathrm{ch}\, qd\,\mathrm{sh}\, sd.$$

In order to compute the above integrals, we analyze them in the complex plane of k. The singularities of the integrands in the plane of k are simple poles, defined by the condition of zero determinants Δ_s and Δ_a. Each real root of the equations $\Delta_s = 0$ and $\Delta_a = 0$ determines the wave number of an excited symmetrical or antisymmetrical Lamb wave. Inasmuch as the receivers of ultrasonic Lamb waves normally respond to surface displacements of the layer (plate), we will be satisfied with finding expressions just for the surface ($z = d$) displacements U_0 and W_0 for $x > v$, i.e., to the right of the excitation zone. Closing the path of integration along the real axis with a semicircle of infinite radius in the upper half-plane of k, we reduce the integrals to the sum of residues of the poles. Thereupon we obtain

$$U_0 = \frac{\sigma_0}{\mu k_t} |\Phi| \left\{ \sum_{k_s} \frac{k_t \sin(k_0 - k)b}{k_0 - k} \cdot \frac{(s^4 - k^4) \operatorname{sh} sd \operatorname{ch} qd\, e^{ikx}}{2k\Delta_s'} + \right. \left. + \sum_{k_a} \frac{k_t \sin(k_0 - k)b}{k_0 - k} \cdot \frac{(s^4 - k^4) \operatorname{sh} qd \operatorname{ch} sd e^{ikx}}{2k\Delta_a'} \right\},$$

$$W_0 = \frac{i\sigma_0}{\mu k_t} |\Phi| \left\{ \sum_{k_s} \frac{k_t \sin(k_0 - k)b}{k_0 - k} \cdot \frac{q(s^2 - k^2) \operatorname{sh} sd \operatorname{sh} qde^{ikx}}{\Delta_s'} + \right. \left. + \sum_{k_a} \frac{k_t \sin(k_0 - k)b}{k_0 - k} \cdot \frac{q(s^2 - k^2) \operatorname{ch} sd \operatorname{ch} qde^{ikx}}{\Delta_a'} \right\}; \tag{1.31}$$

here Δ_s' and Δ_a' are the derivatives of Δ_s, Δ_a with respect to k; and $\sum_{k_s}, \sum_{k_a}$ denote summation over all wave numbers of the possible symmetrical and antisymmetrical normal modes that can exist in a layer of given thickness.

Analyzing the resultant expressions, it is readily seen that they comprise three dimensionless factors (not counting $\sigma_0/\mu k_t$ and e^{ikx}). The first ($|\Phi|$) represents the dependence of the normal stress amplitudes in the excitation zone on the angle θ. The calculation reveals that as θ varies from 0 to 90°, the amplitude of the normal stresses created by a longitudinal wave in the wedge at its boundary with the layer decreases on the average, going to zero at $\theta = 90°$. The decrease, however, does not follow a monotonic course, but undergoes abrupt oscillations, the number of which is greater as the layer becomes thicker. The second factor is associated with the periodic sinusoidal distribution of the driving stresses on the surface of the layer. It represents a rapidly oscillating function (with variation of θ) having alternating maxima and minima. The principal maximum is determined from the conditions $\theta_{s,a} = \sin^{-1} k_{s,a}/k_w$. Its acuity, like that of the neighboring maxima, is greater the larger the dimension of the excitation zone $\bar{b} = k_t b$. The height of the principal maximum is directly proportional to $\bar{b}$. The third factors, which we designate f_{U_0}, f_{W_0}, are determined by the relative thickness of the layer $\bar{d} = k_t d$, the Poisson ratio ν, and the type and order number n of the excited wave. The dependence of the amplitudes of the excited waves on $\bar{d}$, ν, n is very complex. We will analyze it only for limiting cases. For a thin layer ($\bar{d} \to 0$), the function f_{U_0} tends to a constant for the zeroth symmetrical mode s_0, the function $f_{W_0} \to 0$ ($\bar{d}$). This means that for small layer thickness, the zeroth symmetrical mode is excited with an amplitude that is independent of the layer thickness. For the antisymmetrical zeroth mode a_0 in this case, $f_{U_0} \to (\bar{d})^{-1/2}$, $f_{W_0} \to (\bar{d})^{-3/2}$, i.e., the amplitude tends to infinity. For a thick layer, when $\bar{d} \to \infty$, f_{U_0}, and f_{W_0} tend to constants for s_0 and a_0. This implies that for a thick layer, when the modes s_0 and a_0 degenerate into surface Rayleigh waves, their amplitudes no longer depend on the thickness $\bar{d}$. At critical layer thicknesses, with the onset of a new symmetrical or antisymmetrical mode, when $k_{s,a} \to 0$, we have the following: If the mode appear as a longitudinal wave, $f_{W_0} \to k_t/k$, $f_{U_0} \to$ const; if it appears as a transverse wave $f_{U_0} \to$ const, $f_{W_0} \to 0$. Consequently, at the critical layer thicknesses, the amplitudes of the excited longitudinal waves tend to infinity and the amplitudes of the transverse waves tend to a constant value. For large layer thicknesses, when $\bar{d} \to \infty$, for all modes other than s_0, a_0 we find that that $f_{W_0} \to 1/\bar{d}^3$, $f_{U_0} \to 1/\bar{d}^4$, which means that the amplitudes of the excited waves fall off rapidly with increasing layer thickness.

For $x < -b$ (to the left of the excitation zone), the expressions for the surface displacements differ from Eqs. (II.31) in the signs in front of U_0 and ikx in the exponents, and in the second factor, $k_0 - k$ is replaced by $k_0 + k$. With this change of signs, the amplitude of a wave radiated in the $-x$ direction will be considerably less than the amplitude of a wave in the $+x$ direction. For $k_0 = k$ (at the principal maximum), their ratio is of the order $1/k_{s,a}b$, so that the radiation in this case may be regarded as essentially unidirectional.

Figure 40 shows, as an example, the theoretical (1) and experimental (2) dependences of the amplitudes W_0 in the waves s_0, a_0, s_1, a_1, respectively, on the angle θ for a polystyrene wedge on a Dural layer. The theoretical curves are calculated according to Eqs. (II.31). All of the amplitudes are referred to the amplitudes of the principal maxima. The ratio of the amplitudes of the principal maxima and the corresponding ratio for the experimental curves are given below:

Wave mode	s_0	a_0	s_1	a_1
Height of principal maximum				
theoretical curve	1.00	0.87	0.044	0.60
experimental curve	1.0	1.5	0.34	0.58

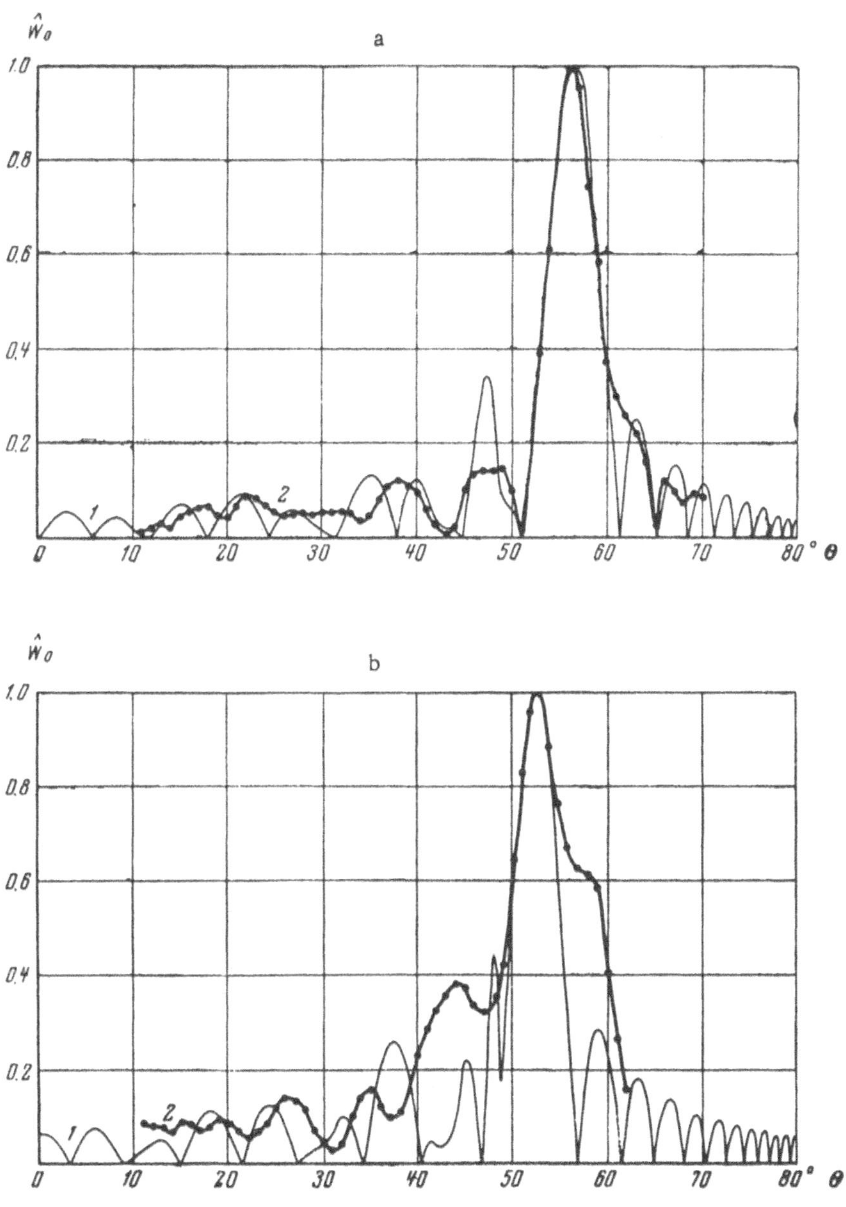
a
$\hat{W}_0$
1.0
0.8
0.6
0.4
0.2
1
2
0
10
20
30
40
50
60
70
80°
θ
b
$\hat{W}_0$
1.0
0.8
0.6
0.4
0.2
1
2
0
10
20
30
40
50
60
70
80°
θ

Fig. 40 (a, b)

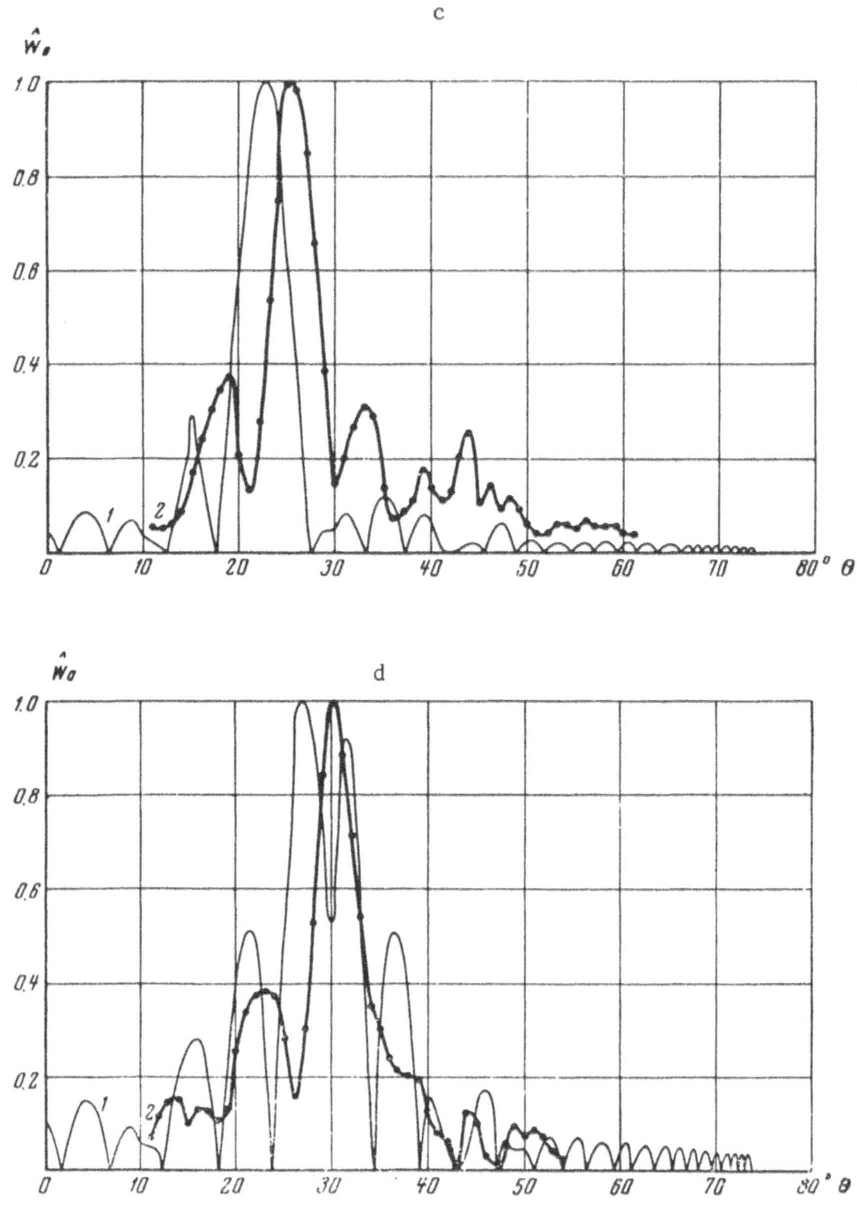
c
$\hat{W}_s$
1.0
0.8
0.6
0.4
0.2
1
2
0
10
20
30
40
50
60
70
80° θ
d
$\hat{W}_a$
1.0
0.8
0.6
0.4
0.2
1
2
0
10
20
30
40
50
60
70
80° θ

Fig. 40 (c, d)

The relative layer thickness $\overline{d}$ was assumed equal to 4.00, the dimension $\overline{b}$ of the excitation zone at $\theta = 0$ was 26.1. For other layer thicknesses, the character of the dependences illustrated remains the same, i.e., the curves for each mode consists of a series of maxima of various heights, between which are zeros, the width of the maxima decreasing with increasing θ and the principal maxima of the curves being located in the vicinity of angles $\theta = \sin^{-1} k_{s,a}/k_w$.

Experiments to record the angular dependence of the amplitudes of Lamb waves and their dependence on the dimension $2\overline{b}$ of the excitation zone, as well as to evaluate the conversion efficiency in the emission and reception of Lamb waves by the wedge method, were carried out in the pulsing regime at a frequency of $\simeq 3$ Mc. The media for propagation of the Lamb waves were Dural strips 0.9 to 3.0 mm thick. The excitation and reception of the waves were accomplished with prismatic transducers having a variable angle. The construction of these transducers is the same as for the Rayleigh-wave experiments (see Section 2, Chapter I).

The piezoelectric elements were 9 x 9 barium titanate plates. All acoustic couplings were established with castor oil films. In operation with a

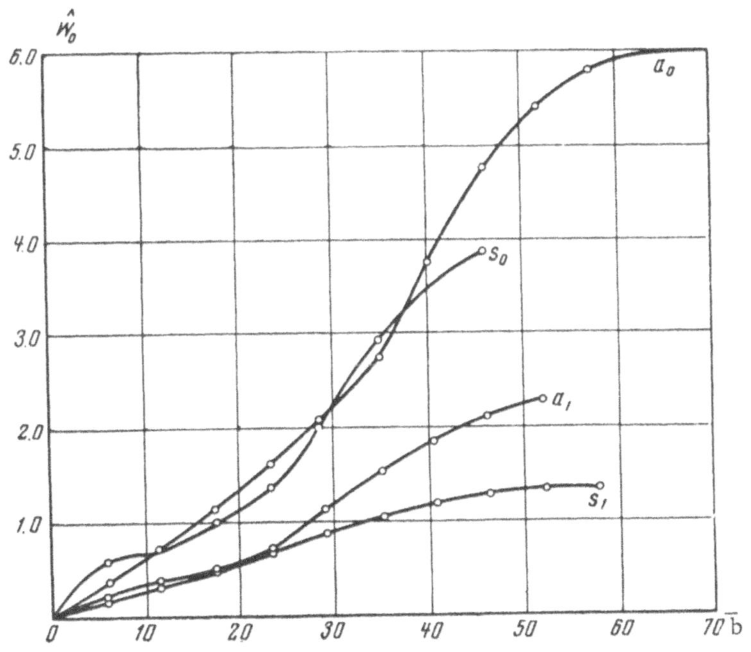

Fig. 41

particular Lamb wave, the receiver was adjusted to the appropriate angle and secured in place. Otherwise, the experimental procedure and equipment were the same as in the Rayleigh wave experiments (see Section 2, Chapter I).

The dots in Fig. 40 (a, b, c, d) represent the experimental dependences $\hat{W}_0(\theta)$ corresponding to the analogous theoretical curves. It is readily seen that the qualitative behavior of the experimental and theoretical dependences is the same. Experimental curves for the dependence of the amplitudes of the waves s_0, a_0, s_1, a_1 on the size of the excitation zone $\bar{b}$ with optimum excitation (at the principal maxima) are shown in Fig. 41. These dependences, as implied by Eqs. (II.31), are described theoretically by straight lines. The experimental curves exhibit distinct "saturation" regions, wherein an increase in the dimension $\bar{b}$ no longer alters the wave amplitude. The quantitative departures of the experimental curves in Figs. 40-41 from the theoretical are explained, clearly, by a lack of complete correspondence between the theoretical idealizations underlying the statement of the problem and the true state of affairs. An analogous situation occurs in the analysis of Rayleigh wave excitation by the wedge method, hence we need not discuss this phenomenon in detail. We conclude this section with Table 5, which shows the double conversion factors K for transformation of the electrical signal into an ultrasonic signal and vice versa, for the modes s_0, a_0, s_1, a_1 and for various plate thicknesses $\bar{d}$. We point out that the losses due to

TABLE 5

Wave mode	$\bar{d}$	$\bar{b}$	K, dB
s_0	0.92	19.7	60.2
a_0		38.6	78.1
s_0		39.8	64.7
a_0		42.4	60.3
s_1	4.00	28.0	78.3
a_1		29.0	61.5
s_0		40.6	67,9
a_0		42.4	67.9
s_1	5.43	28.2	61.2
a_1		31.1	60.4
s_0		38.3	74.9
a_0		38.3	74.9
s_1	8.70	32.3	64.4
a_1		35.1	66.9

attenuation and divergence of the ultrasonic beams in the emitting and receiving prismatic transducers comprise the part of the total conversion losses "included" in K and are equal to 15 dB.

§ 6. Relation Between Lamb and Rayleigh Waves

Lamb waves propagate in plates. Rayleigh waves, strictly speaking, can only propagate over the surface of a half-space. Yet it is impossible in practice, of course, to create a solid half-space. This leads, therefore, to the problem of the possible existence of Rayleigh waves on the surface of solids of finite dimensions. This problem is discussed in [43], which describes an investigation of the structure of waves excited by a source of sinusoidal Rayleigh waves placed on one free surface of a plane infinite plate (solid layer). Inasmuch as only the qualitative aspect of the phenomenon is investigated, the type of emitter is not specified, it merely being assumed that the emitter is capable of exciting Rayleigh waves, i.e., it is, for example, a comb-shaped plate with a spatial period equal to λ_R.

It is shown that with a sufficient layer thickness 2d (d $\sim\lambda_R$ or more, the Rayleigh wave mainly excites in it two normal modes – the zeroth symmetrical and zeroth antisymmetrical modes, due to the similarity of these modes to a Rayleigh wave when $d > \lambda_R$; their phase and group velocities in this case are near the phase velocity of the Rayleigh wave, and the depthwise distribution of the displacements in each of the waves for the upper and lower halves of the layer is similar to the distribution of the displacements in the Rayleigh wave (see Section 4 of the present chapter). The remaining normal modes are excited in an inconsequential degree due to their dissimilarity with the Rayleigh wave. The modes s_0 and a_0 are excited by the emitter with approximately equal amplitudes and phases, since the conditions for their excitation are the same. In the half of the layer where the emitter is located (upper half), the displacements in the modes s_0 and a_0 are in the same direction; in the other half of the layer (lower) they are in opposite directions, since the motion in the mode s_0 is symmetrical relative to the median plane, and in the mode a_0 it is antisymmetrical.

During propagation, the s_0 and a_0 waves interfere with one another. Near the emitter, the phase difference between them is nearly zero, their total acoustic field is similar to the Rayleigh wave acoustic field, hence the described set of waves s_0 and a_0 may be thought of as a quasi-Rayleigh wave. The total distribution of the displacements with depth $(d-z)/\lambda_R$ (measured from the upper boundary of the layer) in the set of waves s_0 and a_0 in phase and in a Rayleigh wave is shown in Fig. 42. The curves are constructed in dimensionless form after the pattern of the curves in Figs. 2 and 33;

$\hat{W}_{s_0}$, $\hat{W}_{a_0}$, $\hat{W}_R$, $\hat{U}_{s_0}$, $\hat{U}_{a_0}$, $\hat{U}_R$ are the displacement amplitudes along the z and x axes in the modes s_0, a_0 and in the Rayleigh wave; $(\hat{W}_{s_0, a_0})_0$, $(\hat{U}_R)_0$ are the corresponding amplitudes for the upper surface. The curves are calculated for a Poisson ratio $\nu = 0.34$, layer thickness $2d = 2\lambda_R$, and with the assumption that $(\hat{W}_{s_0})_0 = (\hat{W}_{a_0})_0$.

On going further away from the emitter, the phase difference between the waves s_0 and a_0 increases, attaining a value of π at some distance. We denote this distance, referred to λ_R, by L. Then the region near the emitter, where the quasi-Rayleigh wave is most like a Rayleigh wave, is defined by the condition $x \ll L\lambda_R$ (here x is the distance from the source). At the distance $L\lambda_R$, a quasi-Rayleigh wave which was originally localized near the surface of the layer where the emitter is situated "transfers over" to the opposite surface. At a distance $2L\lambda_R$, the reverse "transfer" takes place, and so on.

The magnitude of L increases with increasing layer thickness, and for $d/\lambda_R \to \infty$, $L \to \infty$, i.e., the quasi-Rayleigh wave transforms to a Rayleigh wave. The dependence of L on $2d/\lambda_R$ is shown in Fig. 43. For $2d/\lambda_R > 2.5$ the value of L is given, correct to no less than 10%, by the formula

$$L = \left[\frac{1}{4} + \frac{1}{8\left(1-\eta_R^2\right)} + \frac{1}{8\left(1-\eta_R^2\xi^2\right)} - \frac{1}{2-\eta_R^2}\right] e^{2k_R d\sqrt{1-\eta_R^2}}, \tag{II.32}$$

from which it is apparent that the quantity L increases exponentially with the thickness of the layer 2d. For $2d \simeq 2\lambda_R$, $L \simeq 40$; and for $2d \simeq 5\lambda_R$, $L \simeq 60{,}000$.

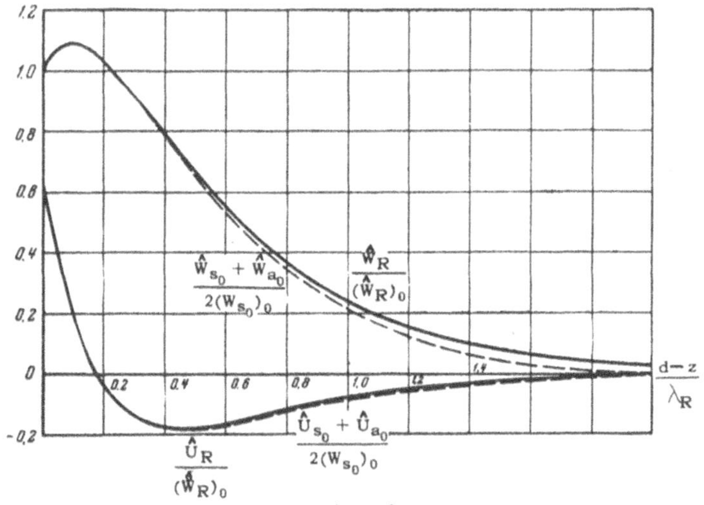

Fig. 42

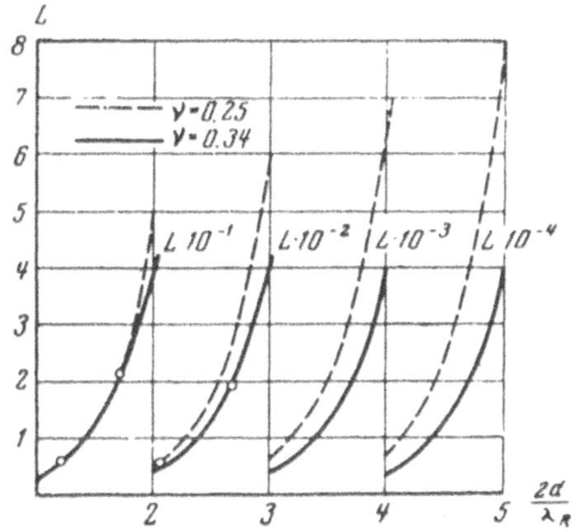

Fig. 43

This means that in metal at a frequency of 3 Mc, the quasi-Rayleigh wave "transfers" from the upper surface of the layer to the lower, covering a distance of about 60 m. At distances from the emitter much less than 60 m (i.e., at least an order of magnitude less), the quasi-Rayleigh wave will be practically indistinguishable from a Rayleigh wave.

The formation and propagation of a quasi-Rayleigh wave in a layer has also been investigated experimentally [43] on a pulsing apparatus consisting of a rectangular pulse generator with a pulse duration of 2-10 μsec and carrier frequency of 2.7 Mc, a prismatic (wedge) emitter and receiver for Rayleigh waves, an amplifier, and an indicator. The experiments were carried out with Dural strips 0.9 to 5 mm thick and fully corroborated the above discussion. A Rayleigh wave was not observed in the strips, but the two zeroth modes, s_0 and a_0 were excited in them. For large distances separating the emitter and receiver, the pulses corresponding to these modes separated, and their group velocities were in good agreement with the analytical values. For small separations, the pulses overlapped and interfered, forming a quasi-Rayleigh wave, which "transferred" from one surface of the strip to the other. The experimentally measured periods of these transfers are marked by dots in Fig. 43. The dots provide a good fit to the corresponding ($\nu = 0.34$) theoretical curves for the dependence of L on $2d/\lambda_R$.

We note in conclusion that the described quasi-Rayleigh wave consisting of two waves of equal amplitude, s_0 and a_0, will not only be excited by a Rayleigh wave emitter placed on one surface of a solid layer (plate) of sufficient thickness, it will also be generated by any distribution of stresses or displacements specified on one of the surfaces of the layer. In this case, other Lamb waves will only be excited in conjunction with a quasi-Rayleigh wave.

§ 7. Directivity Patterns of Lamb and Rayleigh Wave Emitters

In the discussion of Lamb and Rayleigh waves so far, and in particular, the methods of exciting these waves, it has been assumed that the dimensions of the emitters along the wave front (y axis) were infinite and, accordingly, that plane Lamb and Rayleigh waves would propagate in the solid. In practice, however, the dimensions of the emitters along the wave front usually amount to about ten wavelengths. It is reasonable in such cases to be interested in the directivity patterns of the emitters.

In the present section, which is based on the material of [44], we examine this problem for the two most widely employed methods of exciting Lamb and Rayleigh waves: the comb-structure method and the wedge method. In both methods the emitters, situated on the surface of a plate (see Fig. 44) or solid of sufficiently large dimensions (which may be regarded as a half-space), create a perturbation on the surface with a spatial period equal to the zero-order Lamb wavelength or Rayleigh wavelength, respectively.

For calculation of the patterns, we make a number of simplifying assumptions:

1) We regard the solid medium in which propagation takes place as isotropic, homogenous, and perfectly elastic;

2) Inasmuch as the energy of the beam of Lamb or Rayleigh waves diverges in two dimensions, the structure of the waves in the third dimension is completely nonessential for the patterns (Lamb waves of different orders but

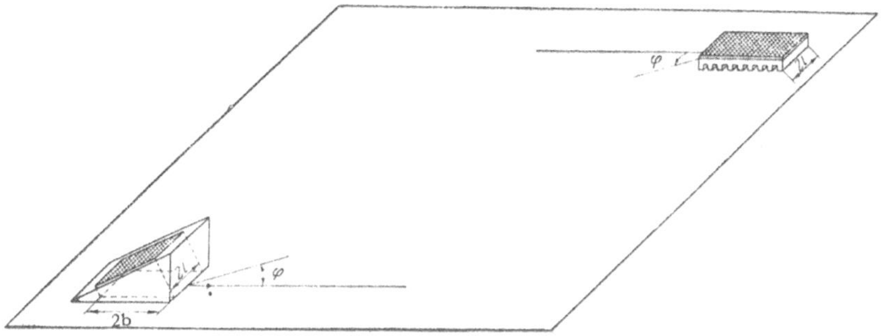

Fig. 44

with the same wavelength will have identical directivity patterns), and we will consider the plane problem, assuming that wave propagation occurs in the plane r, ϕ (Fig. 44);

3) We replace the actual Lamb or Rayleigh wave emitter by a set of elementary emitters appropriately distributed in that region of the plane r, ϕ where the actual emitter is located. In the case of a plate with the comb profile, this region will be a series of strips in the plane r, ϕ which correspond to the projections of the plate; in the case of the wedge it will be the cross section of the beam of longitudinal waves in the wedge (disregarding spreading) in the plane r, ϕ;

4) The amplitude of the displacement created in the plane r, ϕ by the elementary emitter is considered to be proportional to $H_0^{(1)}(kr)e^{-i\omega t-\delta\rho}$, where $k=2\pi/\lambda$ is the wave number of the Lamb or Rayleigh wave. The factor $e^{-\delta\rho}$ accounts for the attenuation of the wave from the elementary emitter in propagation over a path ρ through the region occupied by the other elementary emitters; δ is the attenuation factor per unit path length. This attenuation is an experimentally established fact and is attributed to the transformation of the Lamb and Rayleigh waves into other modes (including Lamb-wave modes of other orders), when the individual surface regions of the layer or half-space are no longer free. For the emitters that we used, the coefficient δ, determined experimentally, was equal to 0.19 mm^{-1} for the comb-shaped plate and 0.20 mm^{-1} for the wedge.

Integrating over all the elementary emitters and carrying out the required calculations, we obtain the following expressions for the directivity patterns of a comb-shaped plate and wedge:

$$N_{cs} = \frac{\sin(kl\sin\varphi)}{kl\sin\varphi}\cdot\frac{\sin\pi\cos\varphi/2}{\cos\varphi}\,e^{\bar{\delta}(1-1/\cos\varphi)}\times$$

$$\times\frac{\sqrt{1+e^{-4n\bar{\delta}/\cos\varphi}-2e^{-2n\bar{\delta}/\cos\varphi}\cos(2\pi n\cos\varphi)}}{\sqrt{1+e^{-4\bar{\delta}/\cos\varphi}-2e^{-2\bar{\delta}/\cos\varphi}\cos(2\pi\cos\varphi)}}\;\frac{1-e^{-2\bar{\delta}}}{1-e^{-2n\bar{\delta}}} \qquad \text{(II.33)}$$

for $\varphi\sim 0$,

$$N_{cs} = \frac{\sin \pi \cos \varphi/2}{kl \cos \varphi} \times$$

$$\times \frac{\sqrt{\sin^2 (kl \sin \varphi)\, \mathrm{ch}^2 4l\bar{\delta}/\lambda \sin \varphi + \cos^2 (kl \sin \varphi)\, \mathrm{sh}^2 4\bar{\delta} l/\lambda \sin \varphi}}{\sin \varphi} \times \qquad \text{(II.34)}$$

$$\times \frac{\sin (n\pi \cos \varphi)}{\sin (\pi \cos \varphi)} \cdot \frac{1 - e^{-2\bar{\delta}}}{1 - e^{-2n\bar{\delta}}} e^{\bar{\delta}(1-4l\,\lambda \sin \varphi)} \qquad \text{for } \varphi \sim \pi/2;$$

$$N_w = \frac{2\bar{\delta}}{\pi} (1 + \mathrm{cth}\, \delta b) \frac{\sin (kl \sin \varphi)}{kl \sin \varphi} e^{-\delta b/|\cos \varphi|} \times$$

$$\times \frac{\sqrt{\sin^2 [kb (1 - \cos \varphi)]\, \mathrm{ch}^2 \frac{\delta b}{\cos \varphi} + \cos^2 [kb (1 - \cos \varphi)]\, \mathrm{sh}^2 \frac{\delta b}{\cos \varphi}}}{\sqrt{(1 - \cos \varphi)^2 + \frac{4\bar{\delta}^2}{\pi^2 \cos^2 \varphi}}}. \qquad \text{(II.35)}$$

Here, N_{cs} and N_W are the ratios of the displacement amplitudes created by the comb-shaped plate and wedge in the ϕ direction in the Fraunhofer zone (i.e., at distances $r >> D^2/2\lambda$, where D is the maximum linear dimension of the emitter along the wave front (cf. [45]) to the corresponding amplitudes for $\phi = 0$; $2l$ and 2b are the linear dimensions of the excitation zones in the plane r, ϕ (Fig. 44); n is the number of projections on the comb-shaped plate; and $\bar{\delta} = \delta\lambda/4$.

Equations (II.33)-(II.35) are only approximate in view of the lack of rigor in accounting for the attenuation of the waves created by the elementary emitters. Equation (II.33) provides a good approximation for $0 < \phi < l/2b$, Eq. (II.34) for $\pi/2 - \lambda/l < \phi < \pi/2$, Eq. (II.35) in the intervals $0 < \phi < l/2b$ and $\pi - (l/2b) < \phi < \pi$. At other angles, the relative error in calculations by these equations amounts to 20-30%, but this is not too significant, since the emission from the comb-shaped plate and wedge at these angles is small.

An experimental investigation of the directivity patterns was performed in the pulsed mode on an apparatus consisting of an electrical rectangular pulse generator with a 2.5-3.5 Mc sinusoidal carrier, an amplifier, and an indicator (type 25I oscilloscope). The pulse duration was either 10 or 2 μsec. A Dural plate with a 19-projection comb profile and a 9 × 9 mm titanate plate mounted on it, as well as a polystyrene wedge with a 55° slope and 9 × 9 mm titanate plate mounted on the sloping face were used as the emitters. The receiver in both cases was a similar polystyrene wedge with a very small acoustic coupling area (circle, 3 mm in diameter) against the surface of the sample in which the waves were propagated. This made it possible to measure the amplitude of surface vibrations on the sample in a small localized region. All acoustic bonds were realized by means of castor oil. To record the directivity patterns, the emitter and receiver were placed on the highly finished surfaces of Dural sheets 45 × 30 cm in area and 6 mm in thickness. The emitter was attached in a fixed position, the receiver was moved around to successive points of a circle 25 cm in radius around the emitter, each measurement of the amplitude at a point being immediately referred to a corresponding measurement on the emitter axis ($\phi = 0$, see Fig. 44). In order to preclude the influence of variations in the acoustic bond between the receiver and Dural sheet on the results of the measurements, every pair of measurements (in the directions ϕ and $\phi = 0$) was repeated twenty times and averaged. The wedge angle and spacing between centers of the comb projections were chosen so that the spatial period of the perturbations on the surface of the Dural sheet would be equal to the Rayleigh wavelength in the Dural. Then for the given sheet thickness, each emitter excited a set of two zero-order Lamb-wave modes, which interfered with one another to create a field similar to a Rayleigh wave field at all points of the sheet (see Section 6).

The directivity patterns of the comb-shaped plate (with $n = 1$ and 19, respectively) are shown in Figs. 45a and 45b, the directivity pattern of the wedge in Fig. 45c. To record the directivity pattern of a comb-shaped plate with one projection, the 19-projection comb was used but with only one projection acoustically coupled with the Dural sheet. The circles in Fig. 45 correspond to a pulse duration of 10 μsec, the crosses to 2 μsec. The theoretical curves (solid lines) were calculated according to Eqs. (II.33)-(II.35). The patterns for the comb-shaped plate were constructed in the interval $0 \leq \phi \leq \pi/2$; for the wedge, in the interval $0 \leq \phi \leq \pi$. The patterns can be obtained outside these intervals by an appropriate symmetrical extension (the symmetry of the empirical patterns was tested and verified experimentally). Analysis of Fig. 45 leads to the following conclusions:

1) The experimentally measured directivity patterns agree only qualitatively with the analytical, the main discrepancy consisting in the almost total absence of fine structure in the experimental diagrams. For the comb-shaped plate with $n = 19$ and the wedge, secondary maxima are generally absent from the experimental patterns (except for slight peaks at $\phi = \pi/2$ for the plate and $\phi = \pi$ for the wedge), while for the comb with one projection they are only slightly pronounced. The experimental patterns suggest the envelopes

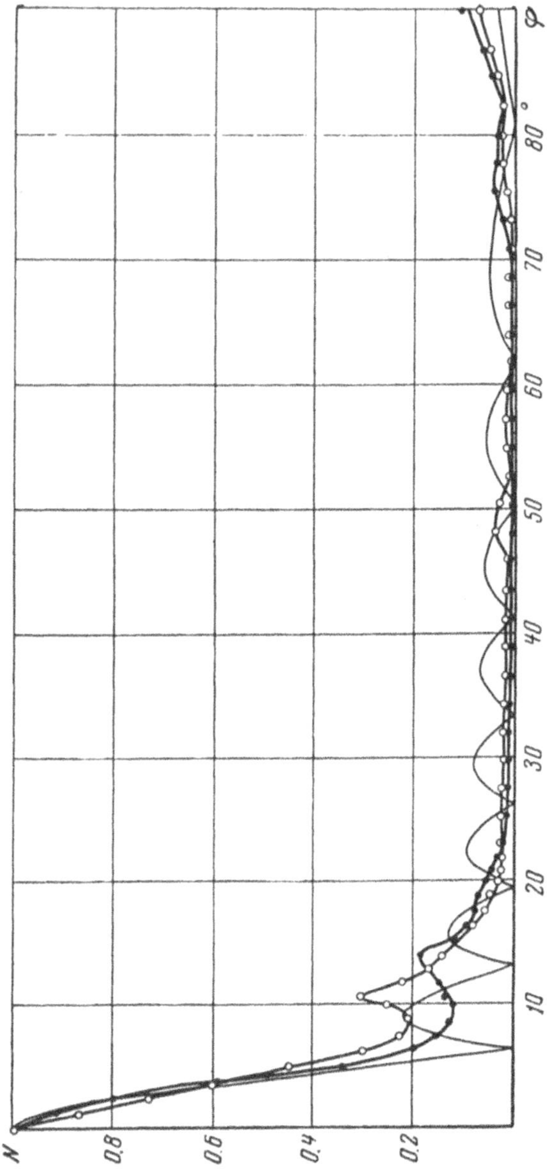

Fig. 45a

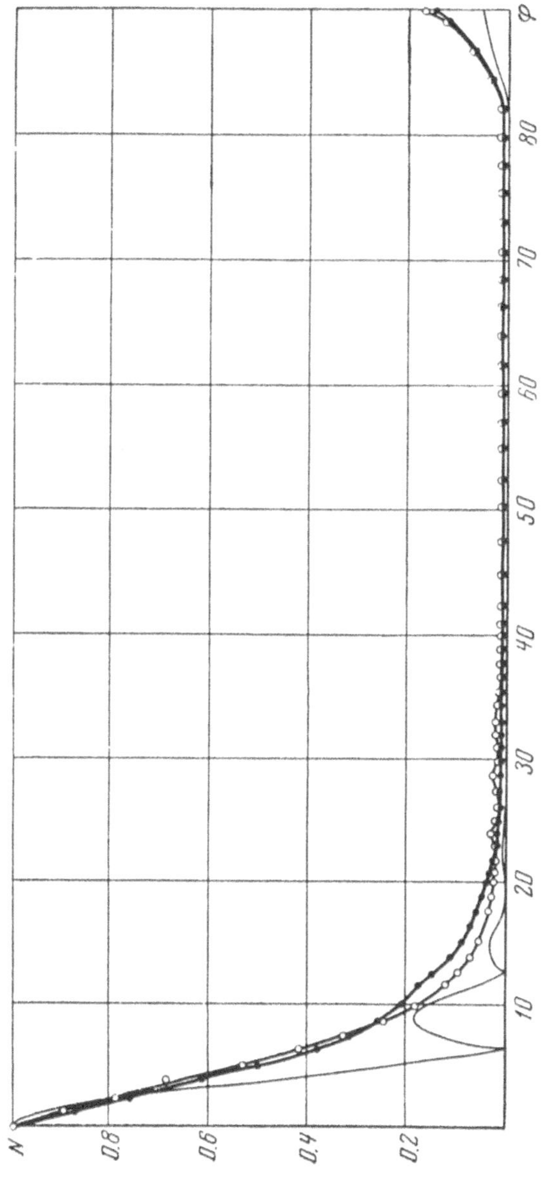

Fig. 45b

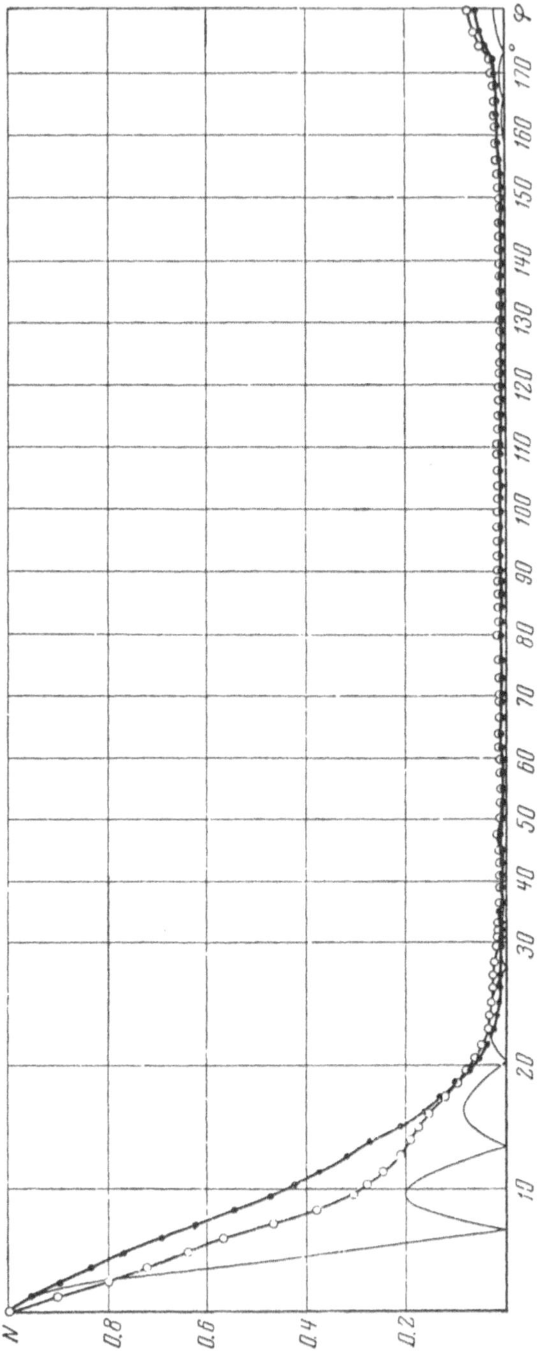
N
0.8
0.6
0.4
0.2
10
20
30
40
50
60
70
80
90
100
110
120
130
140
150
160
170°
φ

Fig. 45c

of the equivalent theoretical curves. Clearly, the main reason for the discrepancy between the experimental and analytical patterns is a lack of correspondence between the idealized theoretical distribution of the displacements in the excitation zone in the plane r, ϕ and the real situation, due to the nonpiston-like vibrations of the titanate plates, nonuniformity of the transition coupling layer, divergence and attenuation of the beam of longitudinal waves in the material of the wedge, and imperfections in the fabrication of the comb-shaped plate (irregular width of the projections and notches, etc.);

2) The angular width of the principal radiation maximum both for the comb-shaped plate with any number of projections and for the wedge is approximately equal to λ/l, i.e., it is determined chiefly by the transverse dimension of the emitters in the plane r, ϕ and does not depend on the second dimension (2b) (the experimental width of the principal maximum for the wedge emitter is somewhat larger, clearly, due to divergence of the beam of longitudinal waves in the wedge). We note that in the case $b \gg l$, which we have not discussed, this situation will no longer be true;

3) As a result of the almost total absence of secondary radiation maxima, the comb-shaped plate may be regarded as a bidirectional emitter, the wedge as a unidirectional emitter;

4) The behavior of the experimental directivity patterns in pulsed operation depends on the pulse duration, but this dependence is only slight (the case of very short pulses with a spatial length much less than the dimensions of the emitter, of course, is excepted). It is perfectly justifiable, therefore, to replace the pulsed mode by continuous operation, as in our calculations.

§ 8. Attenuation of Lamb Waves as a Function of Frequency

The presence of dispersion in the phase and group velocities of Lamb waves has an important effect on the behavior of the attenuation factor for these waves. Whereas the attenuation of waves that do not exhibit phase velocity dispersion does not depend on the dimensions of the sample in which they are propagating and has a fairly smooth dependence on the frequency, for Lamb waves the picture is altogether different. In regions of strong phase velocity dispersion, the attenuation will exhibit a rather steep dependence on the frequency and thickness of the layer (plate), i.e., on $k_t d$. The anomalous behavior of the attenuation factor with strong dispersion on the part of the velocity of a normal mode is an extremely general phenomenon and one that is inherent in normal modes of any kind (acoustic, electromagnetic, etc.) [46]. The dependence of the attenuation factor for Lamb waves on the relative thickness $k_t d$ of the layer in which they are propagating has been investigated in [36, 47]. We will present the fundamental results of these investigations in the present section.

For the calculation of the indicated dependences, we use the method used in calculating the attenuation factor for Rayleigh waves (see Section 3, Chapter I): we specify the attenuation of longitudinal and transverse waves in the solid medium as the imaginary parts of the corresponding wave numbers and derive from the characteristic equations expressions for the imaginary parts of the Lamb waves in terms of the imaginary parts of the longitudinal and transverse wave numbers. Consider a solid layer with losses, where

$$\begin{aligned} \bar{k}_l &= k_l' + ik_l''; \\ \bar{k}_t &= k_t' + ik_t''; \\ \bar{k}_{s_n, a_n} &= k'_{n, a_n} + ik''_{s_n, a_n}. \end{aligned}$$

Let the attenuation of transverse and longitudinal waves be small ($k_l' \gg k_l''$, $k_t' \gg k_t''$). The attenuation of all Lamb waves in this case (as will become apparent later) is also small, except in the vicinity of the critical points. We write the complex wave numbers for the longitudinal, transverse, and n-th-order symmetrical (or antisymmetrical) Lamb waves in the form

$$\begin{aligned} \bar{k}_l &= k_l'(1 + i\alpha); \\ \bar{k}_t &= k_t'(1 + i\beta); \\ \bar{k}_{s_n, a_n} &= k'_{s_n, a_n}(1 + i\gamma_{s_n, a_n}), \end{aligned} \qquad \text{(II.36)}$$

where $\alpha = k_l''/k_l'$, $\beta = k_t''/k_t'$, $\gamma = k''_{s_n, a_n}/k'_{s_n, a_n}$ are small real corrections, numerically equal to the attenuation factors, divided by 2π, for longitudinal, transverse, and Lamb waves per corresponding wavelength. We substitute the complex wave numbers (II.36) into the characteristics equations (II.4), (II.5). Rejecting terms of order α^2, β^2, γ^2 and higher, we separate the equations into their real and imaginary parts. Then the real parts yield equations [agreeing in form with (II.4), (II.5)] determining the phase velocities of the Lamb waves, the imaginary parts yielding equations for γ_{s_n, a_n}. From the latter we obtain the following expressions:

$$\begin{aligned} \gamma_{s_n} &= A_{s_n}\alpha + B_{s_n}\beta; \\ \gamma_{a_n} &= A_{a_n}\alpha + B_{a_n}\beta, \end{aligned} \qquad \text{(II.37)}$$

where

$$A_{s_n}=\frac{\dfrac{\eta_n^2\xi^2}{1-\xi^2\eta_n^2}+\dfrac{2\eta_n\bar{d}}{\sqrt{1-\eta_n^2}\,\mathrm{sh}\,2\bar{d}\sqrt{1/\eta_n^2-\xi^2}}}{C};$$

$$B_{s_n}=\frac{-\dfrac{2\eta_n\bar{d}}{\sqrt{1-\eta_n^2}\,\mathrm{sh}\,2\bar{d}\sqrt{1/\eta_n^2-1}}-\dfrac{4\eta_n^2}{2-\eta_n^2}+\dfrac{\eta_n^2}{1-\eta_n^2}}{C}.$$

Here,

$$C=\left[-\frac{1}{\sqrt{1-\eta_n^2}\,\mathrm{sh}\,2\bar{d}\sqrt{\dfrac{1}{\eta_n^2}-1}}+\right.$$

$$\left.+\frac{1}{\sqrt{1-\xi^2\eta_n^2}\,\mathrm{sh}\,2\bar{d}\sqrt{\dfrac{1}{\eta_n^2}-\xi^2}}\right]\frac{2\bar{d}}{\eta_n}-$$

$$-\frac{8}{2-\eta_n^2}+\frac{1}{1-\eta_n^2}+\frac{1}{1-\eta_n^2\xi^2}+2;$$

$$A_{a_n}=\frac{\dfrac{\xi^2\eta_n^2}{1-\xi^2\eta_n^2}-\dfrac{2\eta_n\bar{d}}{\sqrt{1-\eta_n^2}\,\mathrm{sh}\,2\bar{d}\sqrt{1/\eta_n^2-\xi^2}}}{C_1};$$

$$B_{a_n}=\frac{\dfrac{2\eta_n\bar{d}}{\sqrt{1-\eta_n^2}\,\mathrm{sh}\,2\bar{d}\sqrt{1/\eta_n^2-1}}-\dfrac{4\eta_n^2}{2-\eta_n^2}+\dfrac{\eta_n^2}{1-\eta_n^2}}{C_1};$$

where

$$C_1 = \left[\frac{1}{\sqrt{1-\eta_n^2}\ \mathrm{sh}\ 2\bar{d}\sqrt{1/\eta_n^2-1}} - \frac{1}{\sqrt{1-\xi^2\eta_n^2}\ \mathrm{sh}\ 2\bar{d}\sqrt{1/\eta_n^2-\xi^2}}\right]\frac{2\bar{d}}{\eta_n} - \frac{8}{2-\eta_n^2} + \frac{1}{1-\eta_n^2} + \frac{1}{1-\eta_n^2\xi^2} + 2;$$

$$\bar{d} = k_t' d; \qquad \xi^2 = \frac{c_t^2}{c_l^2} = \frac{k_l'^2}{k_t'^2}; \quad \eta_n = \frac{c_{s_n, a_n}}{c_t} = \frac{k_t'}{k_{s_n, a_n}'}.$$

Consequently, the Lamb wave attenuation factor per wavelength is a linear combination of the analogous attenuation factors for longitudinal and transverse waves.

Analyzing Eqs. (II.37), we are readily convinced that the numerical values of the coefficients A_{s_n} and B_{s_n, a_n} are determined by the type and order number of the Lamb wave, relative layer thickness $\bar{d}$, and Poisson ratio ν. For the zeroth modes in the general case $(0 < \nu < 0.5)$, the attenuation factors γ_{s_0} and γ_{a_0} depend for any thicknesses $\bar{d}$ both on α and on β. As $\nu \to 0$, the "relative contribution" of β to Eqs. (II.37) tends to zero, while for $\nu \to 0.5$ the "relative contribution" of α tends to zero. For $\bar{d} \to \infty$, the attenuation factor for the zeroth Lamb-wave modes tend to the attenuation factors for Rayleigh waves [see Eq. (I.22)].

As Lamb waves of higher than zeroth order make their appearance (in the critical region), the coefficients γ_{s_n, a_n} are determined by α or β, depending on whether the Lamb wave appears as a longitudinal or a transverse wave. For $\bar{d} \to \infty$, the coefficients $\gamma_{s_n, a_n} \to \beta$.

All of the peculiarities mentioned above with regard to the relative contributions of α and β in Eqs. (II.37) are explained by a simple physical consideration: if the Lamb wave consists predominantly of longitudinal components (described by the scalar potential ϕ), then γ_{s_n, a_n} is determined solely by the coefficient β.

The dependence of the coefficients A_{s_n, a_n} and B_{s_n, a_n} on the relative layer thickness has been calculated for $\nu = 0.34$ and is shown in Figs. 46-47.

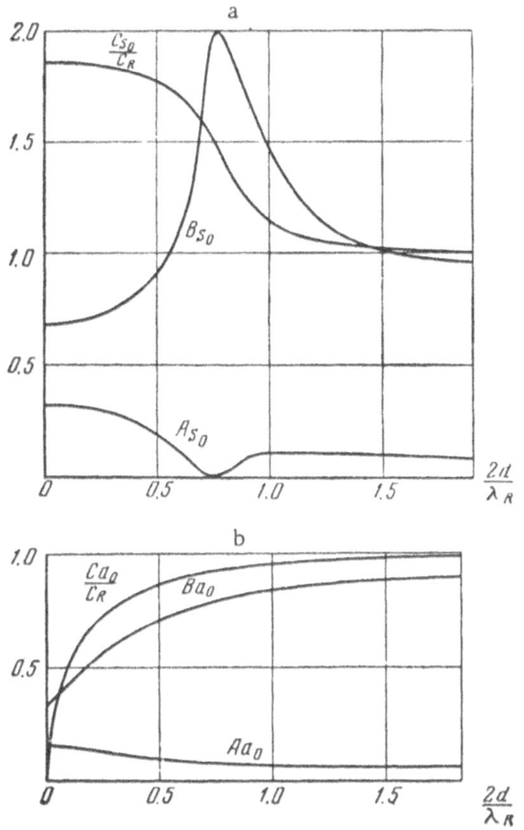

Fig. 46

For the sake of lucidity, the phase velocity dispersion curves of the corresponding normal modes are plotted in the same figures. It is evident from the graphs that, as postulated above, the coefficients A_{s_n, a_n}, and B_{s_n, a_n} depend very strongly on the relative layer thickness, which means that when the attenuation factors α and β for longitudinal and transverse waves are independent of the frequency and layer thickness, the Lamb wave attenuation factor γ depends very strongly on these parameters.

As a rule, the attenuation of Lamb waves is a maximum when the phase velocity dispersion is a maximum. For example, in the critical regions, when the phase velocities and wavelengths tend to infinity, the attenuation factors γ_{s_n, a_n} also tend to infinity. For the s_0 mode in the region of maximum dispersion, the value of B_{s_0} is a maximum, the value of A_{s_0} is a minimum. The maximum of B_{s_0} has a rather large amplitude, so that for the same orders of

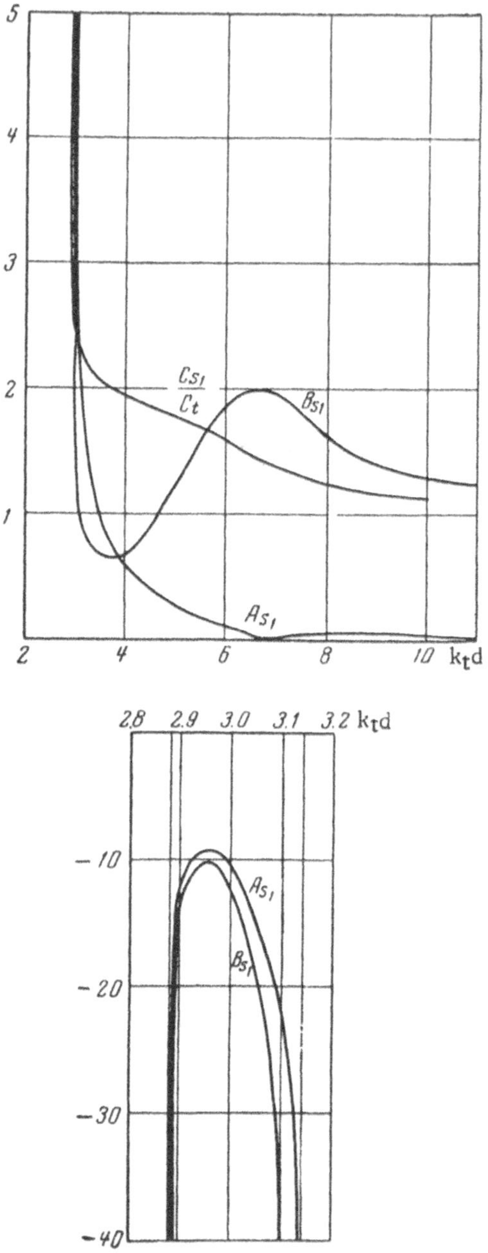
5
4
3
2
1
Cs1
Ct
Bs1
As1
2
4
6
8
10 ktd
2.8 2.9 3.0 3.1 3.2 ktd
-10
-20
-30
-40
As1
Bs1

Fig. 47a

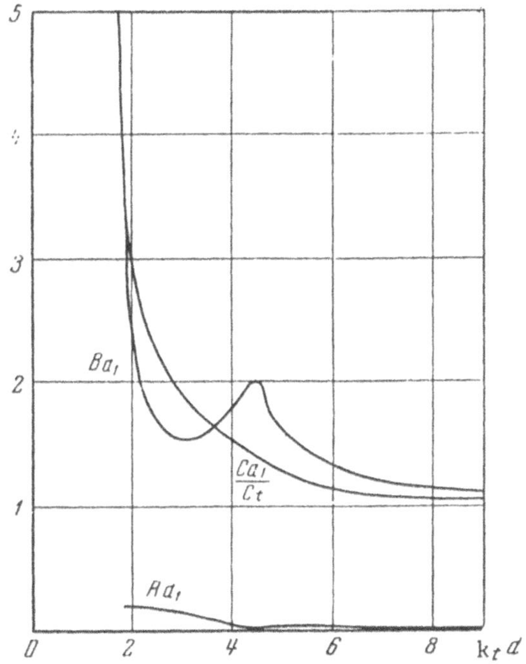

Fig. 47b

magnitude of α and β the region of maximum dispersion will correspond to the maximum attenuation factor γ_{s_0}. Numerically the orders of variation of the coefficients γ_{s_n,a_n} with variation of $\overline{d}$ are the same as the orders of variation of the phase velocities.

Sometimes the coefficients A_{s_n,a_n} and B_{s_n,a_n} become negative – for example, when $k_t d \simeq 3.0$ for the mode s_1 (Fig. 47a). This means that in the given interval the group velocity of the Lamb wave is negative. The results of experiments to test the relations described by Eqs. (II.37) are given in [47]. The experiments were performed at a frequency of 1-2 Mc with samples of low-carbon steel, in which the coefficients γ_{s_0}, γ_{s_1}, γ_{a_0} and γ_{a_1} were measured. The values of the coefficients α in the samples were calculated from these coefficients according to Eqs. (II.37) and compared with the results of the direct measurements of the coefficients α. The agreement between the analytical and measured values of α is not bad, thus confirming the validity of Eq. (II.37).

§ 9. Reflection of Lamb Waves at the End Section of a Plate

In all practical situations, Lamb waves always propagate in plates and strips of finite dimensions. It is reasonable, therefore, that we should be essentially concerned with the problem of their reflection from the edge of a plate or strip. Moreover, insofar as we are aware, this problem has never been treated in the literature.

In the present section, we describe the results of our first experiments (see [48]) on the reflection of the first four ultrasonic Lamb modes (s_0, a_0, s_1, a_1) with the wave at normal incidence on the end section (see Fig. 48, where the x axis is the direction of propagation of the incident wave, the plane $x = 0$ is the end section of the plate).

The experiments were performed in the pulsed mode on an apparatus consisting of an electrical rectangular-pulse generator with a 1-3 Mc sinusoidal carrier, an amplifier, and an indicator (I0-4 oscilloscope). The pulse duration was either 5 or 10 μsec. Lamb waves were propagated into strips 1 to 5 mm thick made of Dural D-16. Their end surfaces from which reflection was to be observed were cut precisely at right angles to the lateral faces. The Lamb waves were generated and received by the "wedge" method described in the fifth section of this chapter. To determine the Lamb-wave reflection coefficient at the end section, we used a procedure completely analogous to the procedure for determining the Rayleigh wave reflection coefficients on dihedral surfaces (see Section 5, Chapter I).

The results of the measurements are shown in Fig. 49. The amplitude reflection coefficient for Lamb waves is plotted on the vertical axis, the quantity $k_t d$ on the horizontal axis. At the present time, there are no solutions available to the problem of Lamb wave reflection from a plate end section for comparison with the experimental data, so we will confine our discussion to the results obtained and an interpretation of the behavior of the reflection coefficients for certain values of $k_t d$.

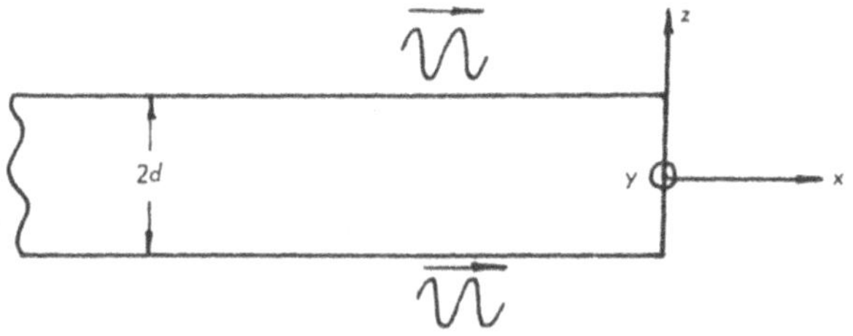

Fig. 48

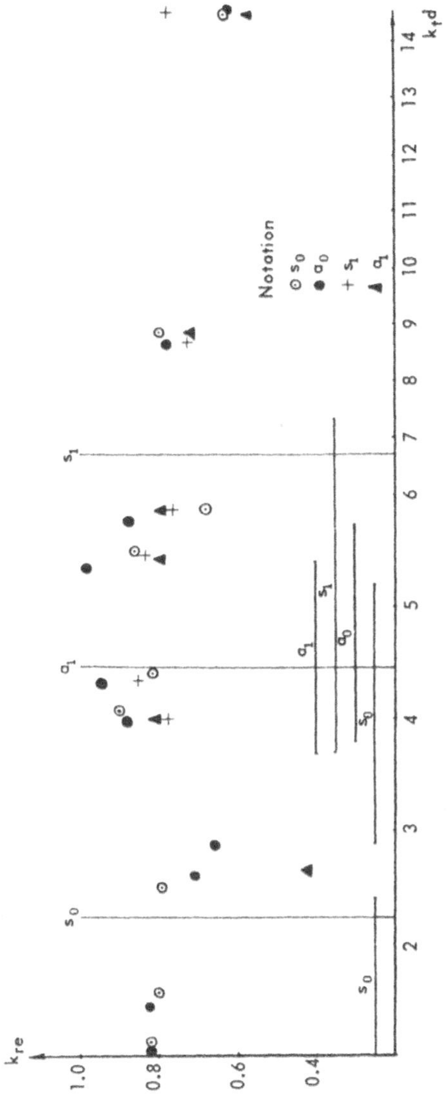

k_{re}
1.0
0.8
0.6
0.4
$k_t d$
2
3
4
5
6
7
8
9
10
11
12
13
14
s_0
a_1
s_1
a_0
Notation
⊙ s_0
● a_0
+ s_1
▲ a_1

Fig. 49.

It is apparent from Fig. 49, that K_{re} depends significantly on $k_t d$, the reflection coefficient being rather large on the average (0.7 or 0.8) at all experimental points $K_{re} < 1$, evincing a partial transformation of the Lamb wave into other modes upon reflection at the end face. The transformation is attributable to the impossibility of the boundary condition of zero stresses σ_{xx} and σ_{zx} at the end surface (Fig. 48) being satisfied by the set comprising the incident and one reflected Lamb wave.

The only exceptions are certain "critical" values of $k_t d$, for which one of the stresses vanishes and, accordingly, K_{re} becomes equal to unity. This occurs in the following situations:

1) for the modes s_0 and a_0 when $k_t d \ll 1$, in which case they represent a longitudinal and a flexural wave in a thin plate, respectively;

2) for all modes except a_0, at those values of $k_t d$ for which the phase velocity of the wave $c = c_t\sqrt{2}$ (these values are marked in Fig. 49 by vertical lines for the modes s_0, s_1, a_1); at these points, unfortunately, experimental values do not exist for K_{re}. For the indicated values of $k_t d$, each Lamb wave represents a set of just two transverse waves traveling at an angle of 90° with respect to one another and at an angle of 45° with respect to the faces of the plate [49];

3) for all Lamb modes except s_0 and a_0 as $k_t d \to \infty$ (practically speaking, for $k_t d > 100$), when the motion therein becomes purely transverse (see Section 4 of the present chapter). The reflection coefficients for the waves s_0 and a_0, which, as described in Section 6 of the present chapter, go over to Rayleigh waves as $k_t d \to \infty$, tend to the value of K_{re} for a Rayleigh wave (cf. the value of K_{re} at $\theta = 90°$ in Fig. 19).

It is reasonable to expect the Lamb wave reflection coefficients to have maxima for those values of $k_t d$ such that one of the stresses in the section $x = 0$ (Fig. 48) is much larger than the other. We chose as a characteristic of the end stress (as a whole for any mode) the integral square of the stress,

$$\int_{-d}^{d} \sigma_{ix}^2 \, dz$$

and calculated the ratio

$$m = \frac{\int_{-d}^{d} \sigma_{xx}^2 \, dz}{\int_{-d}^{d} \sigma_{zx}^2 \, dz}$$

for a Dural layer with thicknesses varying between the limits $k_t d = 0$ to 15. The intervals of values of $k_t d$ in which $m > 10$ or $m < 0.1$ are indicated in Fig. 49 by horizontal lines for each mode. In these intervals, as Fig. 49 shows, the reflection coefficient is substantially larger than for other $k_t d$ (excepting, of course, the above-mentioned cases when $K_{re} = 1$).

§ 10. Propagation in a Cylindrical Layer

In many practical problems (ultrasonic flaw detection, delay lines), Lamb waves are not excited in a plane solid layer, but in a cylindrically curved layer, where the direction of wave propagation is normally perpendicular to the generatrices of the cylindrical surface. A theoretical investigation of the wave propagation in such a layer has been made in [50] by a method analogous to the one we used to investigate Rayleigh waves on cylindrical surfaces.

We will analyze plane harmonic waves (with time dependence given by the factor $e^{-i\omega t}$, which is henceforth omitted for brevity) in a solid cylindrical layer of thickness 2d, these waves propagating in the direction perpendicular to the generatrices, when in cylindrical coordinates r, θ, z the elastic field is independent of z. We may consider the analog of Lamb waves in a plane layer in this case to be the solution of the elasticity equations which has the following properties:

1) satisfies the condition of zero stress on the outer and inner surfaces of the cylindrical layer;

2) depends on the angular coordinate θ according to the law $e^{\pm ip\theta}$, where p is some dimensionless quantity which we will call the angular wave number;

3) goes over to Lamb waves in a plane layer as the radius of curvature of the cylindrical surface tends to infinity, and for finite d and ω.

Under these conditions, the fronts of the waves propagating in the layer will be planes passing through the z axis. The angular wave number p will characterize the phase lead of the wave after traversing a unit angular interval. The relation between the spatial period in the wave for various radii and the number p is determined by the expression $p = (2\pi/\lambda_1)R_1 = (2\pi/\lambda_2)R_2 = (2\pi/\lambda)R$ (see the nomenclature in Fig. 50). We will refer to the wavelength λ in the cylindrical layer as the length of the spatial period in the median surface, the wave number k as the quantity $2\pi/\lambda$, and the phase velocity c as the quantity ω/k. The analogous quantities pertaining to the plane layer will be denoted by the subscript 0 (p_0, k_0, λ_0, etc.). To avoid the need for artificially limiting the problem to the case of integer p, we apply the method of Malyuzhinets [15] and analyze the solution in the infinite angular interval

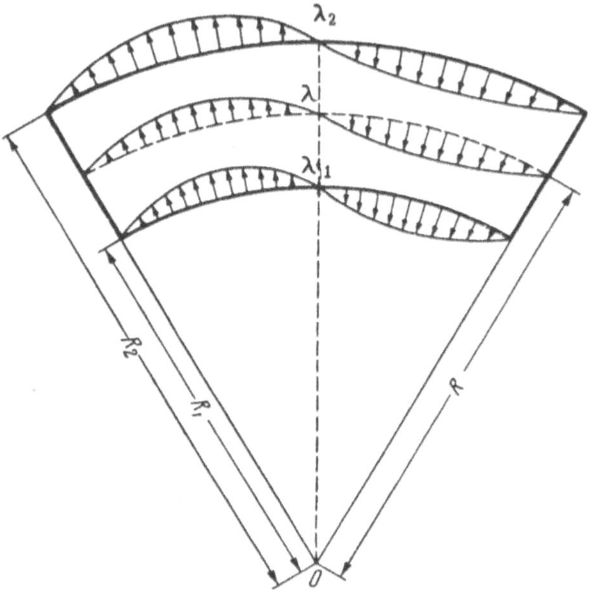

Fig. 50

$-\infty < \theta < \infty$, regarding the z axis as a branch line of infinite order. With this approach, the various values of the solution in the intervals $n < \theta/2\pi < n+1$ for different integer n are logically interpreted as the fields corresponding to successive traversals of the waves around the layer.

We express the solution of the problem in terms of the potentials ϕ and for the longitudinal and transverse waves described by Eqs. (I.23). The connection of ϕ and ψ with the components of the displacements and stresses in the layer is given by the relations (I.24) and (I.25). The general solutions of the wave equations (I.23) have the form

$$\begin{cases} \varphi = [AI_p(k_l r) + CN_p(k_l r)]\, e^{ip\theta}; \\ \psi = [BI_p(k_t r) + DN_p(k_t r)]\, e^{ip\theta}, \end{cases} \tag{II.38}$$

where A, B, C, D are arbitrary constants and $I_p(k_{l,t}r)$, $N_p(k_{l,t}r)$ are p-th-order Bessel and Neumann functions. The conditions of zero stresses σ_{rr} and $\sigma_{\theta r}$ at $r = R_{1,2}$ provide us with four equations for the determination of A, B, C, D. Making use of the identity relations between cylindrical functions, we write

$$
\begin{aligned}
&iaI^-(x_1)A - I^+(y_1)B + iaN^-(x_1)C - N^+(y_1)D = 0;\\
&iaI^-(x_2)A - I^+(y_2)B + iaN^-(x_2)C - N^+(y_2)D = 0;\\
&a[I^+(x_1) - 2bI_p(x_1)]A + iI^-(y_1)B + a[N^+(x_1) -\\
&\qquad - 2bN_p(x_1)]C + iN^-(y_1)D = 0;\\
&a[I^+(x_2) - 2bI_p(x_2)]A + iI^-(y_2)B + a[N^+(x_2) -\\
&\qquad - 2bN_p(x_2)]C + iN^-(y_2)D = 0,
\end{aligned}
\tag{II.39}
$$

where

$$
I^+(z) = I_{p-2}(z) + I_{p+2}(z); \quad I^-(z) = I_{p-2}(z) - I_{p+2}(z);
$$

$$
N^+(z) = N_{p-2}(z) + N_{p+2}(z); \quad N^-(z) = N_{p-2}(z) - N_{p+2}(z);
$$

$$
a = k_l^2/k_t^2 = \xi^2; \quad b = \frac{k_t^2}{k_l^2} - 1; \quad x_{1,2} = k_l R_{1,2}; \quad y_{1,2} = k_t R_{1,2}.
$$

This system has a nontrivial solution when its determinant is set equal to zero. This condition may be specified as follows, after some straightforward transformations:

$$
\begin{vmatrix}
[bI_p(x_1) - I_{p+2}(x_1)] & I_{p+2}(y_1) & [bN_p(x_1) - N_{p+2}(x_1)] & N_{p+2}(y_1)\\
[bI_p(x_1) - I_{p-2}(x_1)] & -I_{p-2}(y_1) & [bN_p(x_1) - N_{p-2}(x_1)] & -N_{p-2}(y_1)\\
[bI_p(x_2) - I_{p+2}(x_2)] & I_{p+2}(y_2) & [bN_p(x_2) - N_{p+2}(x_2)] & N_{p+2}(y_2)\\
[bI_p(x_2) - I_{p-2}(x_2)] & -I_{p-2}(y_2) & (bN_p(x_2) - N_{p-2}(x_2)] & -N_{p-2}(y_2)
\end{vmatrix} = 0. \tag{II.40}
$$

Equation (II.40) is the characteristic equation defining the relationship of the unknown wave number $k = p/R$ to the wave numbers $k_{l,t}$ and the geometry of the layer (R and d). Once k has been determined, three of the four constants A, B, C, D may be expressed in terms of the fourth by means of the system (II.39). Then the expressions (II.38) for the potentials become completely defined, correct to a common amplitude factor.

In order to gain quantitative estimates of the influence exerted by the curvature of the layer on the characteristics of Lamb waves, we analyze the case of a slightly curved layer, when $d/R = k_0 d/p_0 \ll 1$, $\lambda_0/R = 2\pi/p_0 \ll 1$. In this case, adopting $1/p_0$ as the smallness parameter, we can represent the Lamb wave numbers in the cylindrical layer in the form

$$
k = k_0\left[1 + \frac{q(\nu, k_0 d)}{p_0} + \frac{f(\nu, k_0 d)}{p_0^2} + \dots\right], \tag{II.41}
$$

where q and f are certain functions of the Poisson ratio and k_0d, and the cylindrical functions appearing in Eqs. (II.38)-(II.40) may be replaced by asymptotic representations in terms of semiconvergent Debye series [22]. These representations will differ, depending on whether the argument of the function is larger or smaller than the index, which is governed in turn by the thickness of the plane layer corresponding to the given cylindrical layer and by the order number of the Lamb wave in question.

Carrying out the calculations, it can be shown that $q(\nu, k_0, d) \equiv 0$, i.e., to a first approximation the cylindrical curvature does not influence the velocity and other characteristics of Lamb waves in a solid layer.

The calculation of the square correction to k_0 involves very cumbersome operations; we have carried it out only for the two Lamb-wave modes most commonly met in practice—the zeroth symmetrical and zeroth antisymmetrical modes. It is found that for small layer thicknesses, when these modes represent a longitudinal and flexural wave, respectively, in the thin layer, the order of magnitude of $f(\nu, k_0t)$ is equal to unity and $k \sim k_0(1-p_0^{-2})$. For a thick layer ($d \sim \lambda_0$ or more), when both zeroth modes become like Rayleigh waves propagating along the boundaries of the layer, the order of magnitude of $f(\nu, k_0d)$ is equal to $(k_0d)^2e^{k_0d}$, so that $k \sim k_0(1-d^2/R^{2k_0d})$. An exact numerical calculation shows that, given $d/\lambda_0 = 2$, $k = k_0[1-1.61 \cdot 10^6/p_0^2)]$ for $\nu = 0.25$ and $k = k_0[1-(0.98 \cdot 10^6/p_0^2)]$ for $\nu = 0.35$.

Consequently, the correction to the wave number in a cylindrical layer (by comparison with k_0) is a squared quantity with respect to $1/p_0$ and depends significantly on the thickness of the layer and the type and order number of the Lamb wave.

The expression (II.41) may be used to find the phase velocity c of a Lamb wave in a solid cylindrical layer; it is found to be equal to

$$c = c_0\left[1 - \frac{f(\nu, k_0d)}{p_0^2} + \ldots\right].$$

From the familiar relation between the phase and group velocities, we can determine the group velocity of a Lamb wave in a cylindrical layer and show that

$$c^{gr} = \left(c_0 - \lambda_0 \frac{dc_0}{d\lambda_0}\right)\left[\quad + \frac{f(\nu, k_0d) - \dfrac{\partial f(\nu, k_0d)}{\partial (k_0d)} k_0d}{p_0^2} + \ldots\right]. \quad \text{(II.42)}$$

Consequently, the correction to the group velocity due to curvature of the layer is also a squared quantity with respect to $1/\rho_0$ and depends on the layer thickness and type and order number of the Lamb wave.

§ 11. Propagation at an Interface with a Liquid

The problem of Lamb wave propagation in a plate immersed in a liquid is of tremendous practical importance with regard to the immersion technique of testing articles with ultrasonic Lamb waves when the testpiece is immersed in a liquid bath. This problem has been investigated in a number of papers [51-54]. The most detailed, (albeit far from complete) investigation is presented in [51, 54], the basic results of which we now give. By and large, the problem of Lamb waves in a plate surrounded by a liquid is analogous to the problem of Rayleigh waves at the interface between solid and liquid half-spaces, which was analyzed in detail in Section 6 of Chapter I, hence we will make the following presentation as cursory as possible.

Let a plane harmonic wave of frequency ω propagate in the x direction in a solid, perfectly elastic plate of thickness 2d (see Fig. 29), immersed in an ideal liquid. The expressions (II.1) for the wave potentials ϕ and ψ describing the motion of the plate must satisfy Eqs. (I.2), and the expressions for the potential ϕ_L must satisfy the analogous equation (I.43). In accordance with the extinction principle [15], the potential ϕ_L must correspond to waves departing from the plate or to inhomogeneous waves propagating along the faces of the plate and decaying exponentially with distance therefrom. Moreover, in the plane $z = \pm d$, the following boundary conditions must be satisfied: equal displacements in the liquid and plate, pressure in the liquid equal to the stress σ_{zz} in the plate, and zero tangential stress σ_{xz}.

Performing the required calculations, which are for the most part analogous to those carried out in Section 1 of the present chapter, we obtain the same Eqs. (II.8) and (II.9) for the components of the displacements in the plate in the symmetric and antisymmetric modes, except that k_s and k_a in those equations must now be interpreted as the wave numbers of the symmetric and antisymmetric Lamb-wave modes in the plate, taking into account the influence of the liquid. These wave numbers are now determined from the following characteristic equations:

$$(k_s^2 + s_s^2)^2 \operatorname{cth} q_s d - 4k_s^2 q_s s_s \operatorname{cth} s_s d = i\,\frac{\rho_L}{\rho}\,\frac{q_s k_t^4}{\sqrt{k_L^2 - k_s^2}}; \qquad \text{(II.43)}$$

$$(k_a^2 + s_a^2)^2 \operatorname{th} q_a d - 4k_a^2 q_a s_a \operatorname{th} s_a d = i \frac{\rho_L}{\rho} \frac{q_a k_t^4}{\sqrt{k_L^2 - k_s^2}}. \qquad \text{(II.44)}$$

It is readily seen that these equations differ from Eqs. (II.4), (II.5) for a free plate only in the presence of terms on the right-hand sides to account for the influence of the liquid.

It can be shown by investigation of these equations that, in addition to roots which go over to the roots of (II.4), (II.5) as the ratio ρ_L/ρ tends to zero and which correspond, therefore, to "ordinary" Lamb waves in a plate except for the allowance for the influence of the liquid on their characteristics, Eqs. (II.43), (II.44) each have one more real root. One of them corresponds to a symmetrical mode consisting of two inhomogeneous waves in the liquid, which propagate along both faces of the plate and decay exponentially with distance therefrom, plus four inhomogenous waves (two longitudinal and two transverse) in the plate. The other root corresponds to the analogous antisymmetrical mode. The phase velocities of these two waves are smaller than the velocity of the waves in the liquid. The indicated waves are analogous to a surface wave propagating along the boundary between a liquid and solid half-space and consisting of two inhomogeneous waves in the solid half-space and one inhomogeneous wave in the liquid (see Section 6, Chapter I).

The influence of the liquid on a Lamb wave depends very significantly on the ratio of the phase velocities of these waves and the velocity of the waves in the liquid. In essentially every case and for practically all modes except a_0, the condition $c_L < c_{s,a}$ is satisfied. It is clear from physical considerations that in this case the propagation of a wave in the plate will involve the radiation of energy from that wave into the liquid and, accordingly, attenuation along the direction of propagation. This is entirely analogous to the case of a Rayleigh wave on the boundary of a solid and liquid half-space. For an estimate of the attenuation and variation in phase velocity of Lamb waves, calculations of the complex wave numbers $k_{s,a} = k_1 + ik_2$ for symmetrical and antisymmetrical Lamb waves have been made by the author of [54] and by ourselves on the basis of Eqs. (II.43), (II.44). The calculations were made on the assumption of a small ratio ρ_L/ρ. The following relations were obtained for the Lamb-wave attenuation factor per unit transverse wavelength $k_2\lambda_t$ and relative change in phase velocity (due to the influence of the liquid) $\Delta c_{s,a}/c_{s,a}$:

$$k_2\lambda_t = -i \frac{\rho_L}{\rho} k_1 \lambda_t A_{s,a}; \qquad \text{(II.45)}$$

$$\frac{\Delta c_{s,a}}{c_{s,a}} = -B_{s,a}\left(\frac{\rho_L}{\rho}\right)^2. \qquad \text{(II.46)}$$

Here

$$A_s = -\frac{ik_t^4 \operatorname{th} s_s d}{8k_s^2 s_s \sqrt{k_L^2 - k_s^2}}\left[1 + \frac{k_s^2}{2s_s^2} + \frac{k_s^2}{2q_s^2} - \frac{4k_s^2}{k_s^2 + s_s^2} + \right.$$
$$\left. + \frac{k_s^2 d}{2s_s}(\operatorname{th} s_s d - \operatorname{cth} s_s d) - \frac{k_s^2 d}{2q_s}(\operatorname{th} q_s d - \operatorname{cth} q_s d)\right]^{-1};$$

$$A_a = -\frac{ik_t^4 \operatorname{cth} s_a d}{8k_a^2 s_a \sqrt{k_L^2 - k_a^2}}\left[1 + \frac{k_a^2}{2s_a^2} + \frac{k_a^2}{2q_a^2} - \frac{4k_a^2}{k_a^2 + s_a^2} + \right.$$
$$\left. + \frac{k_a^2 d}{2s_a}(\operatorname{cth} s_a d - \operatorname{th} s_a d) - \frac{k_a^2 d}{2q_a}(\operatorname{cth} q_a d - \operatorname{th} q_a d)\right]^{-1};$$

$$B_s = A_s^2\left\{-\frac{1}{2} - \frac{k_s^2}{k_L^2 - k_s^2} + \frac{4iA_s k_s^2 s_s \sqrt{k_L^2 - k_s^2}}{k_t^4 \operatorname{th} s_s d}\left[\frac{2k_s^2}{s_s^2} - \right.\right.$$
$$-\frac{k_s^4}{2s_s^4} - \frac{3k_s^4}{2q_s^4} - \frac{16k_s^4}{(k_s^2 + s_s^2)^2} + \frac{8k_s^4}{q_s^2(k_s^2 + s_s^2)} + \frac{k_s^4 d^2}{s_s^2 \operatorname{sh}^2 s_s d} -$$
$$-\frac{k_s^4 d^2}{q_s^2 \operatorname{sh}^2 q_s d} + \left(\frac{k_s^4 d}{2s_s^3} + \frac{2k_s^2 d}{s_s}\right)(\operatorname{th} s_s d - \operatorname{cth} s_s d) +$$
$$\left.\left. + \left(\frac{3k_s^4 d}{2q_s^3} - \frac{k_s^4 d}{q_s(k_s^2 + s_s^2)}\right)(\operatorname{th} q_s d - \operatorname{cth} q_s d)\right]\right\};$$

$$B_a = A_a^2\left\{-\frac{1}{2} - \frac{k_a^2}{k_L^2 - k_a^2} + \frac{4iA_a k_a^2 s_a \sqrt{k_L^2 - k_a^2}}{k_t^4 \operatorname{cth} s_a d}\left[\frac{2k_a^2}{s_a^2} - \right.\right.$$
$$-\frac{k_a^4}{2s_a^4} - \frac{3k_a^4}{2q_a^4} - \frac{16k_a^4}{(k_a^2 + s_a^2)^2} + \frac{8k_a^4}{q_a^2(k_a^2 + s_a^2)} - \frac{k_a^4 d^2}{s_a^2 \operatorname{ch}^2 s_a d} +$$
$$+\frac{k_a^4 d^2}{q_a^2 \operatorname{ch}^2 q_a d} + \left(\frac{k_a^4 d}{2s_a^3} + \frac{2k_a^2 d}{s_a}\right)(\operatorname{cth} s_a d - \operatorname{th} s_a d) +$$
$$\left.\left. + \left(\frac{3k_a^4 d}{2q_a^3} - \frac{k_a^4 d}{q_a(k_a^2 + s_a^2)}\right)(\operatorname{cth} q_a d - \operatorname{th} q_a d)\right]\right\};$$

the quantities $k_{s,a}$, $s_{s,a}$, $q_{s,a}$ in the equations are interpreted as the corresponding quantities for a plate with free boundaries.

We made some numerical estimates of the attenuation factor $k_2\lambda_t$ and relative variation in velocity for the modes s_0 and a_1 in an aluminum plate immersed in water. For the mode s_0 at small plate thicknesses, when $\bar{d} \to 0$, the attenuation and change in velocity are negligibly small, tending to zero; specifically, $k_2\lambda_t = -0.028\bar{d}$, $\Delta c/c = -0.0005\bar{d}^2$. For a plate thickness $\bar{d} = 2.05$, the attenuation factor for the mode s_0 increases rapidly, becoming equal to -0.232 (the wave decays by a factor of 1/e over a path $\simeq 5\lambda_t$), while the relative variation in velocity amounts to -0.0162, i.e., approximately 1.5%. For a thickness $\bar{d} >> 1$, the attenuation factors and variation in velocity for the modes s_0 and a_0 tend to the corresponding values for a Rayleigh wave (see Section 6, Chapter I). For the mode a_1 near the critical value of $\bar{d}$ when the phase velocity is still rather large and is four times the velocity c_t, the coefficient $k_2\lambda_t$ is equal to -0.047 (the wave is attenuated by 1/e over a path $\simeq 20\lambda_t$), while the relative velocity variation amounts to 0.00011, i.e., $\simeq$0.01%. For a steel plate immersed in water, the values of the attenuation factor for the same values of $\bar{d}$ are approximately one-third lower.

The influence of the liquid on the mode a_0, whose phase velocity (unlike the other modes) tends to zero as $\omega \to 0$ (i.e., for a thin plate) and to c_R as $\omega \to \infty$, has been analyzed in detail in [52] and is summed up in the following. At frequencies below some critical value ω_k (ω_k is somewhat larger than the frequency at which the flexural wave velocity in a free plate is equal to c_L), the influence of the liquid is equivalent to the associated mass effect, and the flexural wave velocity in this case is somewhat smaller than the flexural wave velocity in the free plate. For $\omega > \omega_k$, a second flexural wave can exist in the plate, its propagation involving the radiation of energy from the plate into the liquid. For $\omega = \omega_k$, the phase velocity of the radiating flexural wave is greater than the velocity c_L and the flexural wave velocity in the free plate, and for $\omega \to \infty$ it approaches the latter. For $\omega = \omega_k$, the boundary impedances of the radiating and nonradiating flexural waves are equal, while for $\omega = \omega_k$, the impedance of the radiating wave becomes appreciably less than the impedance of the nonradiating wave, since the radiating wave must be excited by the stronger nonradiating wave.

We note in conclusion that the attenuation of Lamb waves of different orders due to radiation into the liquid strongly depends on the ratio of the vertical to the horizontal component of the surface displacement in the given mode [53]. The attenuation becomes a maximum when the vertical component of the surface displacement is a maximum, and it generally vanishes when the vertical surface displacement is equal to zero, i.e., for example, when $c \rightarrow c_l$, $q \rightarrow 0$ for symmetric modes. This means that if the Lamb wave appears as a longitudinal wave, then in the critical region it decays very rapidly, whereas if the wave appears as a transverse wave; the attenuation, conversely, is very small. For large plate thicknesses, when the surface displacements in the Lamb waves tend to zero by comparison with the volume displacements (see Section 4 of the present chapter), the attenuation due to radiation into the liquid also tends to zero. It is necessary in practice, therefore, in order to achieve long-range propagation of Lamb waves in plates immersed in a liquid, to choose the thickness, frequency, and order number of the Lamb wave such that the ratio $\hat{W}_{s,a}/\hat{U}_{s,a}$ at the plate surface will be minimized as far as possible.

Chapter III

APPLICATIONS OF ULTRASONIC RAYLEIGH AND LAMB WAVES

§ 1. New Engineering Possibilities Afforded by Ultrasonic Rayleigh and Lamb Waves

The use of ultrasonic Rayleigh and Lamb waves in technology has made it possible to solve a number of important practical problems. First and foremost this is true in ultrasonic flaw detection. Previously only two types of waves modes were available for the ultrasonic inspection of materials and products: longitudinal and transverse. However, a required condition for the application of these modes is that all dimensions of the objects to be tested must be much larger than the wavelength. This limitation prevented the ultrasonic inspection of thin-walled materials and structures, as well as the inspection of surface or skin layers of a sample. The latter was attributed to the fact that in the most widely used method of ultrasonic flaw detection – the pulsed method – reflections from surface defects in the sample were inevitably masked by reflections from the surface, due to the limited resolving power of the flaw detector. Consequently, thin-walled parts, surfaces, and surface layers of samples had to be tested by other nondestructive techniques, such as magnetic, X ray, and luminescence testing.

With the advent in ultrasonic flaw detection of Rayleigh waves propagating in the surface layer of a sample, and Lamb waves propagating in thin-walled parts, the above-indicated restrictions on the ultrasonic method of inspection were removed, and it became universalized, i.e., applicable to parts and samples of any shape and of any material. This elevated ultrasonic flaw detection into first place as a method of nondestructive testing.

Tremendous potentials have been uncovered by the application of Rayleigh and Lamb waves for ultrasonic delay lines. Previously, when only volume (longitudinal and transverse) modes were used in such lines, the construction of delay lines was rather complicated, requiring bars or many-angled parallelepipeds of metal, fused quartz, or single crystals. The electrical signal was supplied to the input of the line, was converted into an ultrasonic signal propagating in the acoustic line (bar or parallelepiped), and then (at the output) was converted back into an electrical signal. Any possibility of smoothly varying the signal delay did not exist.

Ultrasonic Rayleigh and Lamb waves have permitted the construction of delay lines with smooth regulations of the signal delay, effected by placing the receiver unit on the surface of the acoustic line and displacing it along that surface. Moreover, due to the localization of Rayleigh and Lamb waves in a thin layer of a solid, it has become possible to reduce the physical dimensions of ultrasonic delay lines significantly; for example, acoustic lines have been produced in the form of metal tapes. Finally – the chief consideration – the application of Lamb waves, which exhibit dispersion in their group velocities, has made it possible to design a conceptually new kind of ultrasonic delay line – a line in which the delay time is frequency-dependent (the so-called dispersive line). Such a line, as will be described in Section 3 of the present chapter, has enhanced the effective range of radar systems, for example.

A third, technically-promising area of application is the use of ultrasonic Rayleigh waves for the comprehensive nondestructive testing of the surface layer of a sample (ascertainment of the degree and depth of thermal anneal, residual stress state, quality of surface finish, etc.). The fact is that the velocity, attenuation, and structure of the Rayleigh wave are directly related to the mechanical, thermal, and other characteristics of the surface layer of a sample in which it is propagating. Consequently, the velocity and attenuation of a Rayleigh wave may be used to gain information regarding the state of the surface layer of a sample.

These three basis regions of application of ultrasonic Rayleigh and Lamb waves will be discussed in somewhat more detail in the ensuing sections.

§ 2. Application in Ultrasonic Flaw Detection

a) Detection of Surface and Near-Surface Defects.

As already noted in the preceding section, Rayleigh waves have opened up a new capability in ultrasonic flaw detection – the inspection of surface and surface-layer defects in materials and finished products. It is a well-known fact that the depth of the localization layer of Rayleigh waves is λ_R to $2\lambda_R$, where λ_R is the Rayleigh wavelength, so that a surface layer of this thickness is accessible for inspection with ultrasonic Rayleigh waves, At the conventional frequencies of 0.5 to 10 Mc used in ultrasonic flaw detection, the approximate thicknesses of a surface layer that can be inspected amount to 0.3 to 6 mm.

It was shown in Section 3, Chapter I that the Rayleigh wave attenuation factor due to absorption and scattering in the material is a linear combination of the corresponding attenuation factors for longitudinal and transverse waves, i.e., the attenuation of Rayleigh surface waves is no greater than the attenuation of volume (longitudinal and transverse) waves. But Rayleigh waves, as opposed to volume waves, do not propagate in the thickness of the material; they propagate along its surface, hence the amplitude of ultrasonic Rayleigh waves decays with distance R due to spreading of the beam as $1/\sqrt{R}$, rather than as $1/R$ as in the case of volume waves. For this reason, ultrasonic Rayleigh waves can propagate over considerably larger distances than longitudinal and transverse waves in the same material. In practice, therefore, solid materials characterized by large or small attenuation of ultrasound are amenable to testing with ultrasonic Rayleigh waves.

In the inspection of samples with Rayleigh waves, it is customary to use the contact version of the the echo-pulse method [6, 7, 54-58], in which defects are disclosed by a pulse (echo) reflected from them. Owing to the ability of Rayleigh waves, as mentioned, to propagate over great distances, in this method of inspection the beam of Rayleigh waves radiated by a source sonically "irradiates" (within the limits of the directivity pattern) a large surface area of the testpiece. It is noted in [7], for example, that a defect 0.025 mm in diameter in a metal beam was reliably detected by means of Rayleigh waves when it was located at a distance of 4 m from the emitter.

The echo-pulse method of inspection can also be realized in the immersion mode, i.e., with immersion of the testpiece in a liquid bath and the emission of ultrasonic waves into the liquid with their subsequent conversion into Rayleigh waves in the piece. This inspection technique does away with the difficulty encountered in establishing a stable and reliable acoustic coupling between the Rayleigh wave emitter (which must be moved over different parts of the test piece during the inspection process) and the test piece itself. The immersion mode is very convenient therefore for the automatic inspection of detailed components. However, as was shown in Section 6 of Chapter I, Rayleigh waves are attenuated at the liquid-solid interface due to the radiation of energy into the liquid, and their effective radius of propagation is about ten wavelengths on the average. For this reason, in the immersion mode of inspection with Rayleigh waves, it is necessary to inspect the component "by parts," moving the emitter systematically about in the bath over all parts of the component. This is obviously why the immersion technique of inspection has not gained widespread usage.

For the excitation of Rayleigh waves in the contact version of the echo-pulse inspection technique, as a rule, the wedge method is used. Typical prismatic search units (probes) for the emission and reception of ultrasonic Rayleigh waves are shown in Fig. 51. The probes in cases a and b, the photographs of which are taken from the monograph [57], have a constant slope and are applicable for the inspection of samples of any one material or of different materials but with the same Rayleigh wave velocity c_R therein. The probe in

Fig. 51

case a has two corrugated surfaces for the scattering of energy from beams of waves reflected from the interface between the probe and sample (i.e., that part of the ultrasonic energy radiated by the piezoelectric plate which has not been transformed into Rayleigh surface wave energy). The probe in case b has a special "spiral" trap for waves reflected from the probe-sample interface, wherein ultrasonic beams are multiply reflected and attenuated. The probe in case c, manufactured by the firm of Lehfeldt [58], is provided with a capability for varying the wedge angle between the limits 25 and 55°, which makes it adaptable to the inspection of a broad category of materials, adjusting in each case to the angle for optimum emission and reception of Rayleigh waves in the tested material.

It was shown in Section 2, Chapter I that the amplitude of an excited Rayleigh wave depends very profoundly on the angle of the emitter wedge, so that the ability to smoothly vary the angle of the wedge transducer greatly enhances the efficiency of this operation, in emission as well as in reception. The angle of inclination of the wedge can be varied by moving the emitting piezoelectric plate (1, in Fig. 52a) around on a special slide block which is acoustically coupled with a circular sector and at the same time can be displaced along it and fixed in any position on it. We used this construction in our investigation of the method for exciting Rayleigh waves (see Section 2, Chapter I). In addition, it is possible to position the emitting piezoelectric

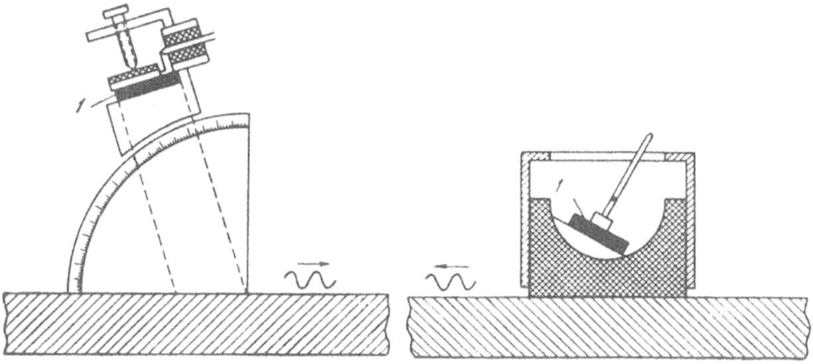

Fig. 52

plate on a cylinder or a cylindrical segment (Fig. 52b), which are inserted in the wedge and form an acoustic bond with it. By rotating the cylinder or segment about the axis, it is possible to vary the inclination of the beam of longitudinal waves impinging on the wedge-sample interface between wide limits. Other constructions of variable-angle wedge transducers are also possible.

For the inspection of defects in the surface layer of products and materials with a low velocity of sound (plastics, rubber, etc.), using Rayleigh waves, it is convenient for the emission and reception of Rayleigh waves to employ the comb-structure method, which is described in detail in Section 2, Chapter I. The comb structure is most simply fabricated as a plate with a ribbed or comb profile, consisting of a periodic alternation of projections and notches, between which the spacing is equal to half the Rayleigh wavelength in the tested material, and an emitting (or receiving) piezoelectric plate acoustically bonded to the comb-shaped plate. The comb-shaped plate is usually made of metal. Photographs of two such plates, made of Dural, are shown in Fig. 51d.

It was shown in Chapter I that Rayleigh waves propagate on a spherical or convex cylindrical surface with the same attenuation as on plane surfaces. The curvature of the surface only changes the phase velocity of the waves. On a concave cylindrical surface, Rayleigh waves are more strongly attenuated than on a plane surface due to the additional radiation of wave energy into the depth of the solid. The attenuation is considerable for small radii of curvature: $R \approx \lambda_R$. For $R > 30\lambda_R$, this attenuation is practically nonexistent.

If two plane surfaces are joined to form a dihedral angle with a sharp apical edge, then for any apex angle θ a Rayleigh wave propagating on one face will pass on to the second face, being partially reflected from the edge. The transmission coefficient, which characterizes the shielding power of the edge, depends on the angle θ and the material of the solid. If the edge between faces is rounded off, the transmission of Rayleigh waves from one face to the other is greatly improved. As the radius of the rounded portion is increased from zero to $2\lambda_R$, the transmission coefficient increases by an oscillating course to unity, while the reflection coefficient, also oscillating, decreases to zero. For a round-off radius $r > 2\lambda_R$, total transmission of Rayleigh waves past the rounded edge takes place.

The above results of the investigations lead to the conclusion that a Rayleigh wave may be propagated essentially over surfaces of any shape, inasmuch as any complex surface may be "resolved" into individual elements consisting of plate surface sections, sections with cylindrical and spherical curvature, and dihedral surfaces with a sharp or rounded edge between.

These results of purely physical investigations are fully utilized and confirmed in practice. Rayleigh waves are used for the inspection of surface and surface-layer defects in samples and components of every possible configuration: rods, beams, springs, wires [7, 10], articulated shafts for tractors [57], solid and hollow circular parts [60, 61], etc. This universality of the shape of samples that can be tested is one of the significant advantages of Rayleigh surface waves over volume waves.

Rayleigh waves can be used to disclose all kinds of surface and surface-layer defects: cracks, surface pitting, voids, segregate, inclusion of foreign impurities, etc. In Section 7, Chapter I, the reflectivities of models of basic surface defects were investigated. It may be concluded from this investigation that on the average a defect reflects Rayleigh waves as an ideal reflector with an average diameter half that of the defect, i.e., very intensely. The reflectivity of near-surface surface defects falls off roughly exponentially with distance from the surface [6]. We mention that the reflection of Rayleigh waves from surface defects is far greater than the reflection of longitudinal and transverse waves from volume defects of the same diameter. This is because the beam of ultrasonic Rayleigh waves always has a "thickness" λ_R to $2\lambda_R$ and, hence, a near-surface defect with a depthwise dimension of λ_R (a crack, for example) intercepts the beam over its entire thickness, creating powerful reflection. A beam of volume waves in any direction is always wider than λ_R or $2\lambda_R$, hence is reflected more weakly from a defect of the same dimension.

Because of the high reflectivity of Rayleigh waves, they are often used for the precision inspection of fine defects, such as fatigue cracks and microcracks in metals [62-64]. Very often this inspection is realized in a continuous mode and permits a crack to be analyzed from the moment of its inception

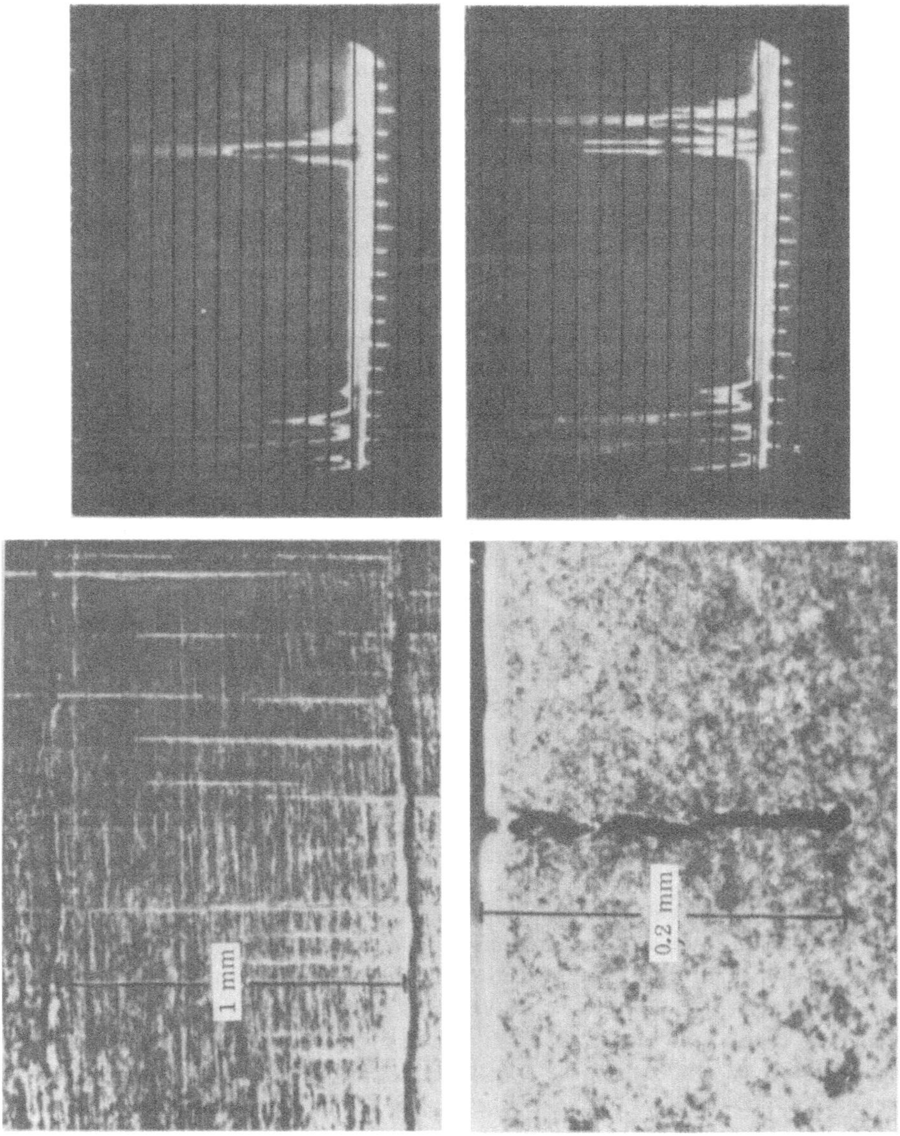

Fig. 53

through its subsequent development. Photographs, taken from [64], of fatigue microcracks on the surface of a steel plate are shown along with oscillograms in Fig. 53. The upper photograph shows a top view (through the microscope, magnified 60 times); the lower photograph shows a transverse section of the crack (magnified 300 times). The upper oscillogram refers to the plate without cracks (having not yet appeared); the initial peaks represent the search pulse, the last peak corresponds to reflection from the edge of the plate. The lower oscillogram corresponds to the same plate, with the formation at its edges of the cracks shown in the photographs. The reflections from them are marked by arrows.

b) Detection of Defects in Sheets and Tubes. While the application of Rayleigh waves in ultrasonic flaw detection has made possible the ultrasonic inspection of surfaces and surface layers of materials and samples, the application of ultrasonic Lamb waves has endowed ultrasonic flaw detection with the ability to inspect sheet materials and structures. Lamb waves are normally used for the inspection of sheet metal, but in general there are no limitations on the material of the sheet, since, as shown in Section 5, Chapter II, Lamb waves are easily excited in a plate of any material, and their range of propagation is considerably larger than the range of propagation of volume waves (like Rayleigh waves, Lamb waves diverge in two dimensions, as a result of which their amplitude falls off due to divergence in proportion to $1/\sqrt{R}$, where R is the distance from the emitter, and not in proportion to 1/R, as is the case with volume waves). The only restriction is that the thickness of the tested sheet must be approximately constant.

At the frequencies normally used in ultrasonic flaw detection practice, in the range 0.5 to 10 Mc, the approximate range of thicknesses of metal sheet that can be tested is from 0.3 to 25 mm. For the inspection of sheet with a thickness less than 1.5 mm, as a rule, the modes s_0 and a_0 are used [40, 65, 66], and for the inspection of thicker sheet the modes s_1, a_1, s_2, a_2, etc. are used [67, 68].

The detection of defects in sheet with Lamb waves is performed using both methods of ultrasonic flaw detection — the echo-pulse method and the shadow method. In testing for defects by the echo-pulse method, the Lamb wave emitter (which is also the receiver (1) in the diagram of Fig. 54) is moved along one edge of the sheet (2). If no defects are encountered between this edge and the opposite one in the path of the ultrasonic beam, only two peaks are seen on the indicator screen (at the beginning and end of the scan), corresponding to the attenuated send-pulse and reflection from the opposite edge. If, on the other hand, the beam encounters a defect (3) in the path from one edge to the other, a third peak will appear between the two just mentioned, due to reflection from the defect (Fig. 54).

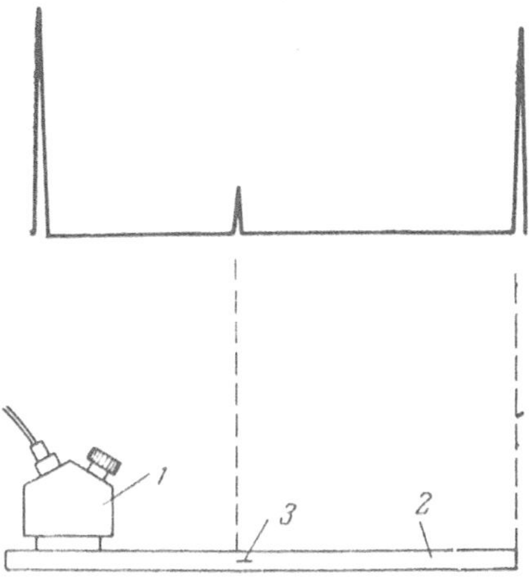

Fig. 54

For inspection by the shadow method, the emitter and receiver are situated on opposite edges of the sheet and are moved synchronously (manually or automatically) in parallel with the edges. When a defect appears in the path between emitter and receiver, the amplitude of the signal on the indicator diminishes. The shadow method is better suited to automatic inspection than the echo-pulse method, but its sensitivity is lower; this means it can only be used to detect extended defects, such as segregate in a sheet material.

Both the shadow and the echo-pulse method of flaw detection in sheet with Lamb waves are used either in the contact or in the immersion mode.

Fig. 55

In the contact mode of inspection, the sheet is placed in the open, while in the immersion mode it is immersed in a water (or other liquid) bath, and the emission of ultrasonic waves takes place at first in the water, with subsequent conversion of these waves into Lamb waves in the sheet.

An automatic system manufactured by the firm of Lehfeldt for the testing of sheet material to be used in building the bodies of small automobiles is shown in Fig. 55. This apparatus is used to test steel sheet 0.8 mm thick by means of Lamb waves for the purpose of disclosing wrinkles and impurities. The inspection is run in the contact version of the echo-pulse method at a rate of 1 m/sec. The sorting device is automatically controlled, whereby the flawless material is automatically segregated from defective material.

The wedge method is used for the emission and reception of Lamb waves in the industrial testing of materials. The wedge emitters (probes) have the same construction as for Rayleigh-wave testing (Fig. 51). Usually a probe with a variable angle is used. The variable angle capability permits just one Lamb-wave mode to be excited predominantly in the tested sample, thus simplifying the inspection operation. Often in both the immersion and contact versions of automatic inspection, a modification of the variable-angle wedge probe is used, wherein the wedge is replaced by a column of liquid between the emitting piezoelectric plate and test piece. A diagram and photograph of such a probe, borrowed from [58], are presented in Fig. 56. The probe is applied to the tested sample from below, the liquid serving two functions in this case: first, it acts as a medium for the propagation of longitudinal waves emitted by the piezoelectric plate; second, it creates acoustic coupling between this medium and the sample. For the purpose of ensuring reliable acoustic coupling, the probe is equipped with a pump which delivers enough liquid to continuously flow over and wet the tested sample at the site of contact with the probe. The circulating liquid is recovered in a circular tank mounted around the probe, to be used over again.

Ultrasonic Lamb waves will reveal the most diverse defects in sheet material: cracks, voids, foreign inclusions, defects in welded joints, segregate, etc. For the inspection of surfaces and surface layers, it is necessary to use the modes s_0 and a_0, in which as shown in Section 4, Chapter II, the motion is localized near the surfaces of the sheet; for the inspection of thickness, higher modes must be used.

Fine cracks and voids, foreign inclusions, and weld defects are disclosed by the echo-pulse method [40, 58, 66]. The detection of nonmetallic inclusions in steel sheet 4 mm thick is described in [66]; inclusions with an area of 1 mm^2 or more are more reliably detected at a radius of 1 m from the emitter. Extended cracks cause considerable amplitude attenuation in the

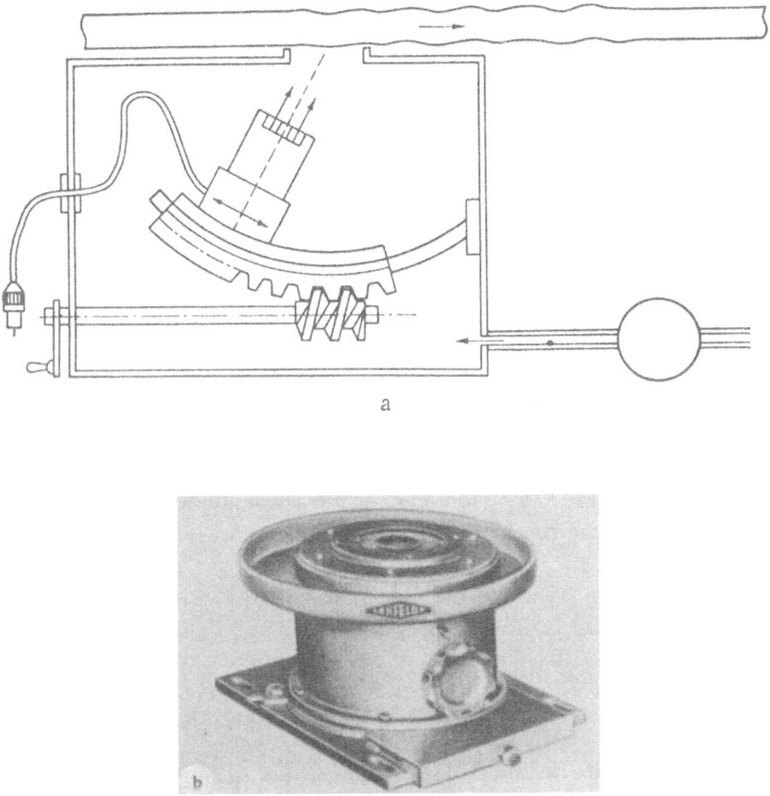

Fig. 56

part of a Lamb wave passing through them, so that they are sometimes searched out by the shadow method. The ability to detect cracks by the shadow method is determined by the size of the crack and the sensitivity of the instrumentation. The attenuating effect of cracks of various depths in the surfaces of a metal plate has been investigated in [69]. Segregate is generally tested by the shadow method.

It is interesting to note that Lamb waves can be used to detect segregate and unbonded zones not only in sheet materials, but even in the surface layer of all kinds of thick samples [41]. A diagram of the scheme for quality control of unbonded flaws between the surface layer and thickness of a metal is illustrated in Fig. 57. The emitter (1) and receiver (2) are placed in the liquid on one side of the test piece (3), where they are separated by an acoustically nontransmissive screen (4) which prevents the transmission of ultrasonic waves directly through the liquid. If there are zones of bond failure in the

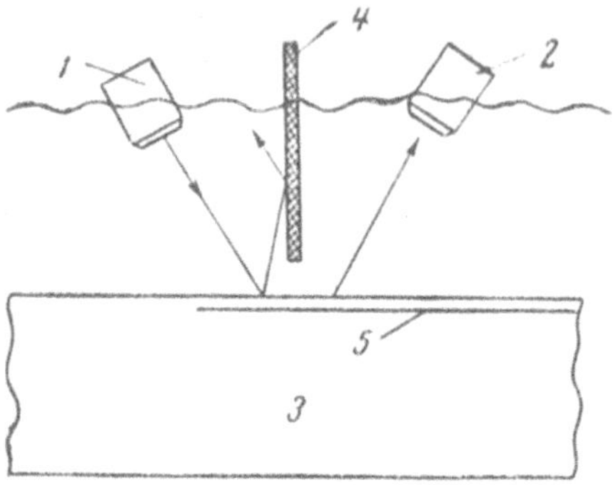

Fig. 57

surface layer, the impingement of longitudinal ultrasonic waves from the liquid onto the separated layer (5) causes Lamb waves to be created therein, and these propagate along the layer, creating a new series of longitudinal waves which are radiated into the liquid and picked up by the receiver. If the bond between the surface layer and the base metal is tight, Lamb waves are not excited and no signal is picked up by the receiver. Consequently, unlike the usual applications of the shadow method, the presence of a defect, instead of causing a reduction, produces the onset of a signal. The same method is used to detect unbonded zones between a thin (approximately 1 mm) aluminum shell and a uranium rod in the fuel cells of atomic reactors [41].

As shown in Section 10, Chapter II, Lamb waves can be generated and propagated in a cylindrical layer of any radius of curvature. Their attenuation with distance in this case is the same as in a plane layer. Lamb waves may therefore be used not only for the inspection of sheet and sheet structures, but tubes as well, where all the principal features and procedures of the testing of sheet carry over to the inspection of tubes. The inspection of tubing with Lamb waves is used very extensively in practice [38, 70, 71]. Usually, the immersion version of the shadow and echo-pulse methods are used for inspection. Lamb waves are propagated around the periphery of the tube perpendicularly to the generatrices of the tube's cylindrical surface. It is possible, by moving the receiver and emitter along the tube, to "scan" the entire tube, section by section.

The exterior view (a) and diagram (b) of the ultrasonic part of a system manufactured at the Central Scientific Research Institute of Technology and Mechanical Engineering (TsNIITMASh) [71] for the continuous factory inspection of tubing is shown in Fig. 58. The apparatus permits the automatic inspection of tubing made from a wide range of materials and alloys, from

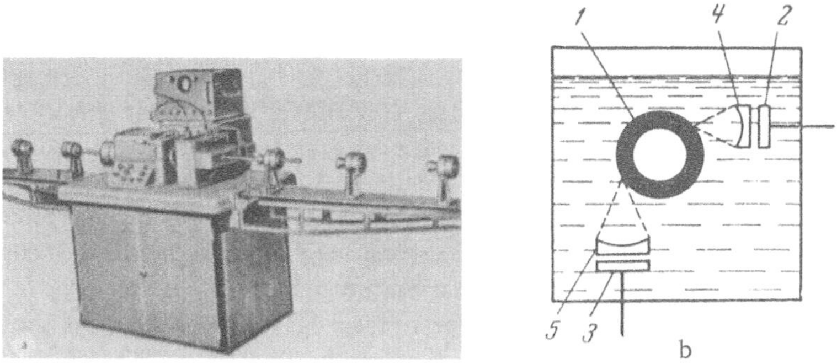

Fig. 58

8 to 60 mm in diameter and 3 m or more in length. The inspection rate is one meter per minute. Defects (cracks, scratches, lap segregate, inclusions, etc.) located in the inside and outside surfaces and in the thickness of the tube wall are disclosed. The section of tubing to be tested (1 in Fig. 58) is placed in a liquid bath in which the search probes (2, 3) with focusing attachments (4, 5) are immersed. Beams of ultrasonic pulses generated by the piezoelectric plates of the probes are transmitted through the liquid layer and are focused onto the surface of the tube, wherein they stimulate Lamb waves. These waves, on being reflected from defects, return to the probes, which act as receiving units in the intervals between sendings. Use of this apparatus permits a speed-up in the inspection rate of three to five times.

c) Detection of Defects in Thin-Walled Articles of Complex Shape. Ultrasonic Lamb and Rayleigh waves may be used to find defects not only in sheets and tubes, but also in thin-walled articles and structures of more complex configuration [72]. Among these are ship hulls (in which it is vitally important to test welded joints), as well as pressed profiles of various ship, aircraft, and automobile structures. The procedures for inspection of these complex parts are the same as for the inspection of sheet and tubing. If the thickness of the part is roughly constant, both types of waves may be used; if the thickness varies from point to point, it is only possible to inspect the surface layers of the part by means of Rayleigh waves.

The application of Lamb and Rayleigh waves has found particularly widespread use for the inspection of gas-turbine blading, combustor casings, and other thin-walled aircraft components [39, 73, 74]. This inspection process is performed without dismantling the engine (in situ at the airfield), which means considerable efficiency and economy and raises the reliability of engine operation. In [73, 74], a procedure is described for this type of inspection, along with the construction of a small-scale portable flaw detector designed especially for the purpose.

The inspection of turbine blading (Fig. 59) is carried out chiefly with emphasis on the edges, where the onset of cracks and other flaws is the most hazardous. The inspection process is realized with ultrasonic Rayleigh waves in the contact version of the echo-pulse method. The emitting and receiving probe (2) has a special step-notched contacting surface to facilitate coupling with the web of the blade (1). Two operators run the inspection. One of them, lying in the jet pipe, applies the probe to the edges of the blade, the second observes the behavior of the oscillograms on the flaw detector display screen (3). A thin film of oil is deposited on the contacting surface of the probe, and the latter is applied to the exit edge at the root section, such that the ultrasonic waves are directed toward the other end of the blade (Fig. 59). The appearance of an echo signal between the initial a and final b (corresponding to reflection from the blade tip) pulses on the flaw detector screen indicates the presence of a defect, which may be through cracks 0.5 mm or more in depth starting from the edge of the blade and running across the web of the blade, microcracks in the surface layer of the blade material, corrosion flaws in the material, deep scratches or dents in the edge of the blade, tiny internal subsurface flaws, etc. Experience has demonstrated the high efficiency of the ultrasonic method of inspecting turbine blading, as evinced by the sharp drop in incidents of blade failure in the engine flow-through section.

The ultrasonic inspection of combustor casings in aeroengines is accomplished with Lamb waves in the contact version of the echo-pulse method. The combustor casing (Fig. 60) represents a thin walled tube (1) with a wall thickness of 1.2 mm. Fatigue cracks, which ultimately result in damage to

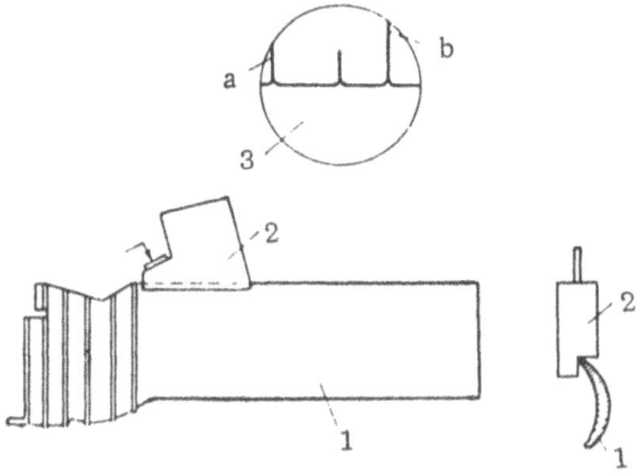

Fig. 59

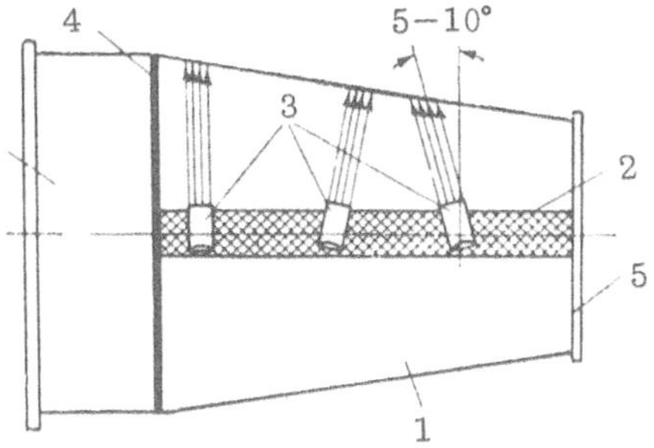

Fig. 60

the casing, are usually formed in a section having the shape of a truncated cone. For their detection, a thin film of oil is deposited on the exterior surface of the casing in the form of strips directed along the generatrices (2). A wedge probe (3) is applied to the casing so that ultrasonic Lamb waves will be directed around the periphery of the casing, then during the inspection process is moved along the lubricated coupling strip (three probe positions are indicated in Fig. 60). During inspection, the probe is rotated from t to 10° from the initial position, thus improving the defect search scan. When a defect is present in the casing, a pulse appears on the flaw detector display screen. There is no final pulse in this inspection technique, so that in order to check the working order of the probe it is necessary from time to time to aim it at a welded joint (4) or seal ring (5) of the casing, so as to obtain a reflected signal from one of those objects. Tests have shown that the described inspection technique enables one to discover fatigue cracks, as well as scratches, lap, nonmetallic inclusions, and other flaws.

§ 3. Application in Ultrasonic Delay Lines

It was noted in Section 1 of the present chapter that the application of ultrasonic Rayleigh and Lamb waves in delay lines has made it possible to design new types of these devices – namely, dispersive, branched delay lines and lines with smooth regulation of the delay time. Moreover, owing to the localization of these waves in a layer rather than in the volume of a solid, delay lines using Rayleigh and Lamb waves are considerably more economical in size by comparison with ultrasonic delay lines using volume modes.

The simplest construction of an ultrasonic delay line using Rayleigh waves with a wide range of delay times is described in [75]. The line is a hollow aluminum cylinder with a helical track on its exterior surface (Fig. 61). The outside diameter of the cylinder is 15.2 cm, its length 61 cm, the

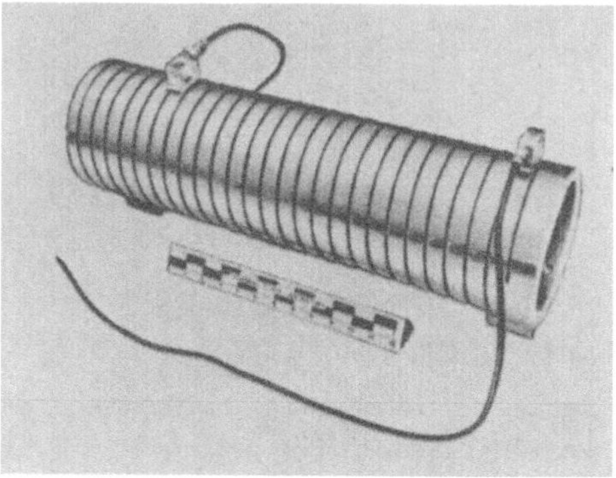

Fig. 61

remaining dimensions are seen in Fig. 61 (in inches). Rayleigh waves are propagated along the helical track. The emission and reception of the waves are accomplished with plexiglass wedge transducers, with the emitting transducer attached in a fixed position, the receiving unit capable of being moved to any position on the helical track. As a result, the delay time may be smoothly varied from a few microseconds to 6 msec. The transducers are acoustically coupled to the surface of the cylinder with an oil film. At a frequency of 0.9 Mc and a relative transmission bandwidth of 15%, for a delay time of 100 μsec the double conversion coefficient for transformation of an electrical pulse into an acoustic pulse and back again amounts to approximately 30 dB. The internal loss coefficient in the line due to absorption and scattering of ultrasonic Rayleigh waves in aluminum is 1.5 dB per msec delay time. The level of spurious signals is approximately 30 dB lower than the basic signal level.

Lamb waves are normally used for the design of dispersive ultrasonic delay lines. Particular interest attaches to so-called linear ultrasonic delay lines, in which the delay time depends linearly on the frequency of the Lamb waves propagating in them. Such delay lines are used extensively in radar focusing devices [76, 77], i.e., systems for increasing the effective radar range without any loss in resolving power. In these systems, a short pilot pulse is decomposed by means of a dispersive delay line into the frequency spectrum, each component of which is shifted in time relative to another. Each component is amplified, and as a result the entire pulse is radiated as a long frequency-modulated wave train of high power. At the receiver, the signal passes through another line with linear dispersion but with the opposite slope, which restores the pulse to its original short duration. In this way, the radar resolving power is not diminished, but the pulse power is singificantly enhanced.

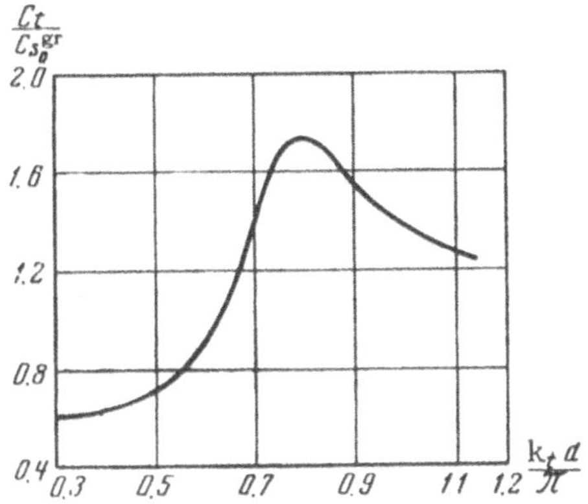

Fig. 62

Usually the zeroth symmetrical mode s_0, propagating in a metal strip, is used for the design of a linear dispersive delay line. The mode s_0 has a linear interval in the frequency dependence of its dealy time. The dependence of $c_t/c_{s_0}^{gr}$ (which is proportional to the delay time per unit path) on $k_t d/\pi = 2df/c_t$, where f is the ultrasonic frequency, is shown for illustration in Fig. 62, which is taken from [78]. It is evident from Fig. 62 that as $k_t d/\pi$ varies in the interval from 0.6 to 0.8 the curve has an approximately linear behavior.

The construction of a linear dispersive ultrasonic delay line using a pulse of the mode s_0 as the propagating ultrasonic signal is described in detail in [78, 79]. The line is an aluminum strip with a thickness of 2.2 mm and length of 142 cm. At one end of the strip (Fig. 63) is placed a rectangular emitting plate (1) of barium titanate ceramic, at the other is an analogous receiving plate. The emitting plate, which is positioned symmetrically with respect to the central plane of the strip, excites a pulse in the mode s_0 in the strip, along which it then propagates. The excitation frequency is 1 Mc; the width of the transmission band for the entire system is 0.1 Mc. The total delay time of the signal produced by the line is equal to 633 μsec, the variation in delay time with varying frequency between the limits of the linear section amounts to 200 μsec with a ±3 μsec deviation from linearity. The average losses due to absorption and scattering of ultrasound in aluminum do not exceed 15 dB. To reduce the level of spurious signals occurring due to spreading of the ultrasonic beam in the line, as well as to obtain a level transmission band, one of the side edges of the strip and certain parts of the base surface are bonded with a special coating (2) that absorbs ultrasonic vibrations. As a result of this, the spurious signal level is 40 dB lower than the basic signal level.

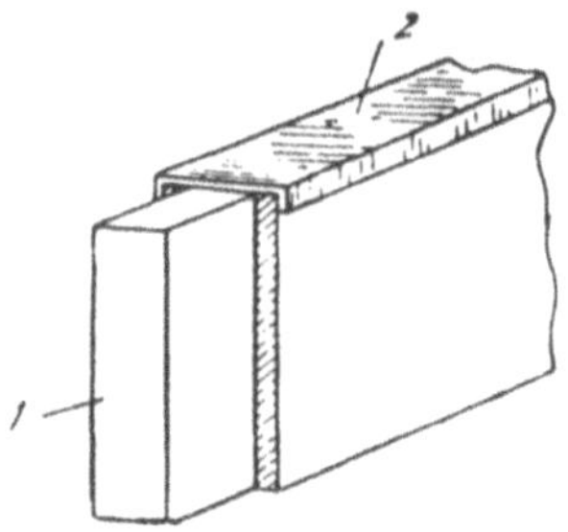

Fig. 63

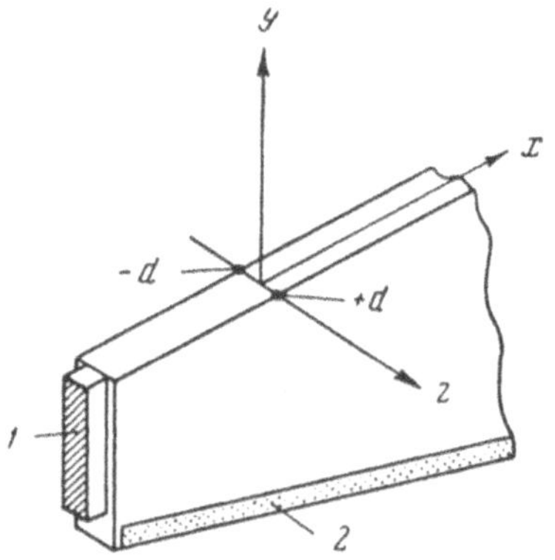

Fig. 64

The width of the line may be fairly arbitrary: within an interval of widths 20 to 40 mm the delay time is almost independent of the width of the line. The thickness of the line must be maintained constant with fairly strict precision (to 10^{-4}d), otherwise the ultrasonic attenuation in the line will be greatly increased due to the nonuniform thickness.

In order to reduce the internal losses and the spurious signal level [807], as well as to improve the transmission bandwidth, the dispersive strip delay line in the mode s_0 is sometimes made in the form of a tapered strip, which broadens in the direction from the emitter to the receiver and is coated with an absorbent material on the tapered face (2 in Fig. 64], 1 is the piezoelectric emitter plate). The thickness 2d of the line is chosen in this case so that the mid-frequency of the transmission band of the line in the mode s_0 satisfies the condition of zero stress, not only in the planes $z = \pm d$ (this is always fulfilled) but also on the face $y = 0$. Then the losses in the line at the mid-frequency are minimized, since reflection of the s_0 wave from the face $y = 0$

(and the associated energy losses from the basic signal) is absent. Simultaneous with the proper choice of thickness, the Poisson ratio of the material of the line is chosen so that the transmission band of the line will fall in the interval wherein the frequency dependence of the delay time is linear. Consequently, it turns out that the geometry of the line may be regulated, and not only the dispersive properties of the line but also the transmission band may be predetermined. The sloping cross section and absorber deposited on it minimize the spurious signals associated with reflection from the faces of the strip and make the shape of the transmission curve smooth.

The parameters of a number of lines of the types described, with a mid-frequency of 2 Mc, are given in [78]. For example, a line made of aluminum strip with a thickness of 0.111 cm, length of 241.3 cm, and width of 0.927 cm at the narrow end, 1.35 cm at the wide end has a transmission band between the limits of 1.92 Mc, the losses in conversion and attenuation of the waves in the line come to 34 dB, and the variation of the delay time for different frequencies within the limits of the transmission band is from 100 to 1240 μsec.

By varying the thickness of the strip delay line in the mode s_0 in the direction of wave propagation, it is possible to obtain a variety of delay characteristics (i.e., the dependence of the delay time on the frequency and transmission bandwidth) which are not obtainable with strips of constant thickness. A strip delay line whose thickness is varied in some manner along its axis may be interpreted as a set of strip lines of constant thickness. It is possible by compounding the delay times of these lines to obtain an over-all delay characteristic that is completely different in shape from the delay characteristics of the individual sections. This possibility was first described and realized in [81], which presents several examples of strip delay lines of variable thickness, both with linear and nonlinear frequency dependences on the part of the delay time.

Strip delay lines using Lamb waves are very simple to make in the form of a variable tapped line; all that needs to be done, instead of picking up vibrations from the end of the line, is to take them from one of its fundamental surfaces (the planes $z = \pm d$ in Fig. 64). The receiving units in this case may be a set of wedge transducers or plates with a comb profile, setting them at various distances from the emitter.

In the investigation of the dispersion curves for the phase and group velocities of Lamb waves of various orders, it was noted (see Section 3, Chapter II) that in certain frequency intervals the phase and group velocities of symmetrical Lamb-wave modes vary extremely little. The frequency intervals corresponding to quasiconstant values of the phase and group velocities are of interest with regard to the distortion-free transmission of signals. Considering the presence of a whole series of such frequency intervals, it may be stated that a solid plate (strip) can, in principle, serve as a dispersionless multichannel delay line and transmission line. The notion of putting a solid plate to this use was advanced in [34].

We observe in conclusion that strip delay lines using ultrasonic Lamb waves may be nondispersive as well as dispersive. This is easily effected by using the same frequencies and strip thicknesses at which Lamb waves have negligibly small group velocity dispersion, i.e., by taking, for example, the frequency intervals corresponding to quasiconstant values of the phase and group velocities, as mentioned above. For a nondispersive delay line, it is simplest again to use the mode s_0, making the thickness of the strip much less than the wavelength of this mode at the required frequency, whereupon phase and group velocity dispersion is practically nonexistent (see Figs. 32a and 32b).

§ 4. Other Applications

Lamb and Rayleigh waves are used in quite a large number of useful and important applications in addition to the two very widespread major areas described above.

Foremost among these is the application of ultrasonic Rayleigh waves, not just for the detection of flaws in the surface layer of a sample, but for the overall analysis of surfaces and surface layers. This testing is normally carried out by precision measurements of the velocity and attenuation of ultrasonic Rayleigh waves, since both of these fundamental characteristics of a Rayleigh wave are very sensitive to variations in the mechanical parameters of a surface layer in which they are propagating. For example, the attenuation of a Rayleigh wave gives an indication of the quality of surface finish of a sample [31], provided that all of the other parameters (chemical composition, mechanical structure, etc.) of the tested series of samples are identical. In this case, the attenuation is uniquely related to the degree of surface finish. The velocity of a Rayleigh wave can be used to determine the stresses in the surface layer of a part [82, 83], because, all other conditions being equal, the velocity of a Rayleigh wave is uniquely related to the magnitude of the stresses in the surface layer. It is remarked in [82] that the residual stresses in surface layers of metal samples after their machining alter the Rayleigh wave velocity by multiples of ten percent. This variation is fairly easy to detect. It is proposed in [84] that the velocity of Rayleigh waves be used to acquire information regarding the thermal properties of a sample.

It is possible from the attenuation of ultrasonic Rayleigh waves also to determine the depth of a surface-hardened (e.g., case-hardened or annealed) layer of a component. This can be done on the basis of the following considerations. The density and elastic constants of a surface-hardened layer differ somewhat from the same parameters in the base thickness of the sample. Despite this disparity, it is still possible for a Rayleigh wave to propagate in such an inhomogeneous solid medium, although certainly this is a more complex type of wave than in a uniform solid half-space. The attenuation of this Rayleigh wave is determined by the depth of the hardened layer, the densities and elastic constants of the layer and of the base thickness of the material [85]. If all of the parameters except the layer depth are known, the wave attenuation may be used to determine that depth. The practical procedure is very simple [83]. Let there be, for example, a large series of uniform parts with a scatter of values for the depth of the surface-hardened layer. Several parts (with Rayleigh wave attenuation factors as dissimilar as possible) are sampled as reference standards, and the Rayleigh wave attenuation is measured in them. Then, some kind of decision technique is applied to determine the layer depth in the standards. By constructing a curve for the dependence of the layer depth on the Rayleigh wave attenuation, it is possible to ascertain the depth of the layer for any part from the series without destroying it, just by measuring the attenuation; the method can be simplified so that only the amplitude of the preceding signal need be measured with a constant separation between emitter and receiver.

An important practical application of Lamb waves is their use for precision measurements of the thickness of sheet structures which are only accessible from one side. The measurements are carried out in this case in the immersion mode of the echo-pulse method, employing the following procedure.

First, a calibration nomogram is made up for the material of the test batch. For this, a set of plates of various thicknesses and made of the same material is picked out. The plates are placed in a water bath, such that one face is in contact with the water, the second is exposed to the air. In the water are placed a wideband emitter and wideband receiver, which are inclined at a fixed angle θ with respect to the plane of the test plate. A rectangular electrical pulse with a carrier frequency that can be varied within the limits of the system transmission band is delivered to the emitter. For certain fixed (for a given plate) frequencies, the spatial periods of the perturbations generated by the wave from the emitter on the surface of the plate coincide with the Lamb wavelengths in the plate (at these frequencies). This is equivalent to the effective excitation of Lamb waves (see Section 5, Chapter II), and sharp signal peaks appear on the display screen of the indicator. Each maximum corresponds then to a definite Lamb-wave mode. By measuring the frequencies at which the peaks occur, it is possible to obtain for plates of different thicknesses a series of curves representing the dependence of the peaking frequency on the thickness of the plate. Each curve will correspond to its own Lamb-wave mode. With a family of such calibration curves (nomogram) available, it is then possible to determine the unknown thicknesses of samples by placing them in the water bath and measuring the frequencies corresponding to the signal peaks shown on the indicator. The relative position of these frequencies enables one to determine the order of the Lamb-wave mode corresponding to each frequency. Once this is done, the unknown thickness is read off at once from the nomogram.

It is shown in [87] that at frequencies of 2 to 13 Mc the range of thickness of a metal article that are measurable by the indicated procedure extends from 0.015 mm to several millimeters, the precision of the measurements falling within an error of 0.0003 mm, i.e., greatly in excess of the precision attainable with other nondestructive measurement techniques. The immersion mode enhances the speed of thickness measurements with Lamb waves and makes it possible to automate the testing process.

The phase velocity of Lamb waves propagating in a sample with a known thickness and density can be used to calculate the elastic constants of the material—Young's modulus and the shear modulus. This method of determining the elastic constants is very useful when the geometry of the sample renders the use of volume (longitudinal and transverse) waves for the same purpose prohibitive.

These, in brief, are the principal applications of ultrasonic Lamb and Rayleigh waves as described in the literature. The area of these applications is currently in a state of continual expansion.

LITERATURE CITED

1. Lord Rayleigh, On waves propagated along the plane surfaces of an elastic solid, Proc. London Math. Soc., 17:4-11 (1885).
2. E. A. Naryshkina, General theory of Rayleigh waves for a half-space, Trudy Seismologicheskogo Instituta Akad. Nauk SSSR, No. 90 (1940).
3. H. Lamb, On waves in an elastic plate, Proc. Roy. Soc. (London), Ser. A, 93:114 (1917).
4. H. Kolsky, Stress Waves in Solids, Part I, Chapt. II, Sec. 5, Moscow (1955).
5. L. Bergmann, Ultraschall und seine Anwendung in Wissenschaft und Technik, Edwards, Ann Arbor, Mich.
6. E. G. Cook and H. E. Valkenburg, Surface waves at ultrasonic frequencies, Am. Soc. Testing. Mater. Bull., 198:81-84 (1954).
7. C. Minton, Inspection of metals with ultrasonic surface waves, Nondestructive Testing, 12(4):13-16 (1954).
8. F. Firestone and I. Frederick, Refinements in supersonic reflectoscopy, J. Acoust. Soc. Am., 18(1):200-211 (1946).
9. K. N. Vinogradov and G. K. Ul'yanov, Measurements of the velocity and attenuation of ultrasonic surface waves in hard materials, Akust. Zh., 5(3):290-293 (1959).
10. D. S. Shraiber, Ultrasonic flaw detection, Collection: Flaw Detection in Metals, Gos. Izd. Oboron. Prom., Moscow (1959), pp. 241-355.
11. L. C. Lynnworth, Shear wave probes and applications, Inst. Electronics. Elect. Engrs. Trans., Sonics and Ultrasonics, SU (1964).
12. A. G. Sokolinskii, Technique for the Excitation and Reception of Surface Waves, Author's Certificate No. 19297 (1958).
13. I. A. Viktorov, Investigation of methods for exciting Rayleigh waves, Akust. Zh., 7(3):295-306 (1961).
14. E. A. Naryshkina, Vibrations of a half-space under any initial conditions, Trudy Seismologicheskogo Instituta Akad. Nauk SSSR, No. 45 (1934).
15. G. D. Malyuzhinets, Some Generalizations of the Method of Images in the Theory of Sinusoidal Wave Diffraction (Doctoral Thesis), Fiz. Inst. in P. N. Lebedev, Akad. Nauk SSSR (1950).

16. F. Press and I. Healy, Absorption of Rayleigh waves in low-loss media, J. Appl. Phys., 28(11):1323-1325 (1957).
17. I. A. Viktorov, Certain Aspects of the Propagation of Rayleigh Waves in Solids (Masters Thesis), Akust. Inst. Akad. Nauk SSSR, Moscow (1958).
18. I. A. Viktorov, Damping of surface and spatial ultrasonic waves, Akust. Zh., 10(1):116-118 (1964).
19. W. M. Ewing, W. S. Jardetzky, and F. Press, Elastic Waves in Layered Media, McGraw-Hill, New York (1957).
20. I. A. Viktorov, Rayleigh-type waves on a cylindrical surface, Akust. Zh., 4(2):131-136 (1958).
21. I. A. Viktorov, Attenuation of Rayleigh waves on cylindrical surfaces, Akust. Zh., 7(1):21-25 (1961).
22. G. N. Watson, Theory of Bessel Functions, Cambridge Univ. Press, New York (1962).
23. I. A. Viktorov, Influence of surface imperfections on the propagation of Rayleigh waves, Dokl. Akad. Nauk SSSR, 119(3):463-465 (1958).
24. I. A. Viktorov, Transmission and reflection of Rayleigh waves by rounded edges of various radii, Akust. Zh., 7(1):90-91 (1961).
25. G. I. Petrashen', The Rayleigh problem for surfaces waves in the case of a sphere, Dokl. Akad. Nauk SSSR, 52(9):763-766 (1946).
26. I. Cl. de Bremaecker, Transmission and reflection of Rayleigh waves at corners, Geophysics, 23(2):253-266 (1958).
27. V. G. Gogoladze, Rayleigh waves at the boundary between a compressible fluid medium and a solid elastic half-space, Trudy Seismologicheskogo Instituta Akad. Nauk SSSR, No. 127 (1948).
28. L. M. Brekhovskikh, Waves in Layered Media, Chapt. I, Izd. Akad. Nauk SSSR, Moscow (1957).
29. I. A. Viktorov, E. K. Grishchenko, and T. M. Kaekina, An investigation of the propagation of ultrasonic surface waves at the boundary between a solid and a liquid, Akust. Zh., 9(2):162-170 (1963).
30. L. M. Brekhovskikh, The propagation of surface Rayleigh waves along the uneven surface of an elastic solid, Akust. Zh., 5(3):282-289 (1959).
31. N. S. Bykov and Yu. G. Shneider, An experimental investigation of the effect of surface quality on the attenuation of surface waves, Akust. Zh., 6(4):501-503 (1960).
32. I. A. Viktorov, Influence of surface defects on the propagation of Rayleigh waves, Collection: Application of Ultrasonic Vibrations for the Investigation of the Properties, Quality Control, and Machining of Metals and Alloys, Akad. Nauk Izd. UkrSSR, Kiev (1960), pp. 54-61.
33. I. A. Viktorov and T. M. Kaekina, The scattering of ultrasonic Rayleigh waves at models of surface defects, Akust. Zh., 10(1):30-33 (1964).

34. E. S. Mashashvili, Investigation of Normal Modes in a Flat Plate, Report of the Acoustics Institute, Academy of Sciences of the USSR, Moscow (1962).
35. L. V. Agafnova, Ural-1 Computer Program for the Calculation of the Phase and Group Velocities of Symmetrical Normal Modes Propagating in a Solid Plate, Report of the Acoustics Institute, Academy of Sciences of the USSR, Moscow (1961).
36. I. A. Viktorov, O. M. Zubova, and T. M. Kaekina, Investigation of Lamb Waves and the Limits of Their Application for Ultrasonic Flaw Detection, Report of the Acoustics Institute, Academy of Sciences of the USSR, Moscow (1961).
37. R. H. Lyon, Response of an elastic plate to localized driving forces, J. Acoust. Soc. Am., 27(2):259-265 (1955).
38. A. S. Matveev and M. F. Krakovyak, An ultrasonic free-wave flaw-detection method for inspection of thin-walled tubes, Zav. Lab., 11:1235-1238 (1960).
39. I. G. Rasmussen, Ultrasonic inspection of turbine and compressor rotor blades for cracks and other flaws, Nondestructive Testing, 16(3):228-236 (1958).
40. H. Hentze and H. Wiegand, Fehlerortung in Feinblechen durch Ultraschall-Plattenwellen [Flaw detection in metal sheet with ultrasonic plate waves], Materialprufung, 2(2):45-50 (1960).
41. D. C. Worlton, Ultrasonic testing with Lamb waves, Nondestructive Testing, 15(4):218-222 (1957).
42. I. A. Viktorov, O. M. Zubova, and T. M. Kaekina, Investigation of Rayleigh wave excitation by the "wedge" method, Akust. Zh., 10(4):412-418 (1964).
43. I. A. Viktorov and R. A. Grigoryan, Quasi-Rayleigh waves in an elastic layer, Akust. Zh., 5(3):366-368 (1959).
44. I. A. Viktorov and O. M. Zubova, Directivity diagrams of radiators of Lamb and Rayleigh waves, Akust. Zh., 9(2):171-175 (1963).
45. G. S. Gorelik, Vibrations and Waves, Chapt. 8, Gostekhteorizdat, Moscow (1950).
46. V. L. Ginzburg, On the general connections between the absorption and dispersion of sound waves, Akust. Zh., 1(1):31-39 (1955).
47. K. V. Zharkov, L. G. Merkulov, and E. D. Pigulevskii, Normal mode damping in a plate with free edges, Akust. Zh., 10(2):163-166 (1964).
48. I. A. Viktorov, Reflection of ultrasonic Lamb and Rayleigh waves by a rectangular section of a plate, Akust. Zh., 12 (1966) (in preparation).
49. W. P. Mason, Physical Acoustics, Vol. I, Part A, Chapt. II, Academic Press, New York-London (1964).

50. I. A. Viktorov and O. M. Zubova, Normal plate modes in a solid cylindrical layer, Akust. Zh., 9(1):19-22 (1963).

51. A. Schock, Der Schalldurchgang durch Platten [Sound transmission in plates], Acustica, 2(1):1-17 (1952).

52. G. Kurtze and R. H. Bolt, On the interaction between plate bending waves and their radiation load, Akust. Beih., 9(1):238-242 (1959).

53. D. C. Worlton, Experimental confirmation of Lamb waves at megacycle frequencies, J. Appl. Phys., 32(6):967-971 (1961).

54. L. G. Merkulov, Damping of normal modes in a plate immersed in a liquid, Akust. Zh., 10(2):206-212 (1964).

55. G. I. Binczewski, Standardization and application of ultrasonic surface-wave inspection, Nondestructive Testing, 15(1):36-40 (1957).

56. I. Tabin and M. Kurek, Badanie metali ultradzwiekowymi falami powierzchniowymi [Testing of metals with ultrasonic surface waves], Hutnik, 23(7-8):300-304 (1956).

57. L. Filipczynski, Z. Pawlowski, and I. Wehr, Ultradzwiekowe metody badania materialow [Ultrasonic Methods of Testing Materials], Panstwowe Wydawnictwa Techn., Warsaw (1959).

58. W. Lehfeldt, Testing of sheet material, tubes, and bars with ultrasonic surface and plate waves, Proceedings of the International Symposium on the Application of Ultrasonics [Russian translation], Bratislava, Czechoslovakia, September, 1962.

59. E. Haeusler, Application of elastic (ultrasonic) surface waves to nondestructive testing of solid media, Materialprufung, 2(2):51-55 (1960).

60. W. Grabendörfer, Untersuchung von dünnen runden Werkstoffen mit Überschall-Oberlächenwellen [The examination of thin round objects by means of ultrasonic surface waves], Stahl und Eisen, 75(19):1273-1275 (1955).

61. W. Grabendörfer, The Examination of Thin Round Objects by means of Ultrasonic Surface Waves, Ministry of Aviation, Great Britain, February (1960).

62. P. I. Brosens, N. A. G. Hakimi, and G. R. Knabbaz, Detection of Fatigue Damage with Rayleigh Waves, Applied Mechanics Laboratory, MIT, Cambridge, Mass. August (1960).

63. I. G. Rasmussen, Prediction of fatigue failure using ultrasonic surface waves, Nondestructive Testing, 20(2):103-110 (1962).

64. R. Pohlman, Untersuchungen zum Nachweis von Oberflächenrissen an nitrierten Vergügsstählen mittels Ultraschall-Oberfächenwellen [Detection of surface cracks in nitrated heat-treated steels by means of ultrasonic surface waves], Z. Instrumentenk., 71(12):322-329 (1963).

65. W. Lehfeldt, Die Fehlererkennbarkeit bei der Ultraschallprüfung

mittels Plattenwellen [Flaw detectability in ultrasonic testing with plate waves], Industrieblatt, 60(1):1-7 (1960).

66. C. Drouet, Sondage ultrasonore des tôles minces d'acier à l'aide des ondes de Lamb [Ultrasonic inspection of thin sheet steel by means of Lamb waves], Métaux (Corrosion-Ind.), 37(440):127-153 (1962).

67. P. Höller and E. Lechky, Prüfung von 3 bis 20 mm dicken Stahlblechen mit Ultraschall-Plattenwellen [Testing of 3 to 20 mm thick steel sheet with ultrasonic plate waves], Arch. Eisenhüttenw., 33(10):699-710 (1962).

68. W. Lehfeldt, Ultrasonic testing of sheet material with Lamb waves, Mater. Testing, 4(9):331-337 (1962).

69. T. N. Grigsby and E. J. Tajchman, Properties of Lamb waves relevant to the ultrasonic inspection of thin plates, Inst. Radio Engr. Trans. Ultrasonics Eng., UE-8(1):26-33, March (1961).

70. R. Pohlman, Die kontinuierliche Ultraschallprüfung langsgeschweister Rohre [Continuous ultrasonic testing of longitudinally welded pipe], Bleck, 4(6):25-28 (1957).

71. A. S. Matveev and M. F. Krakovyak, New apparatus for ultrasonic flaw detection in thin-walled tubing, Trudy Tsentr. Nauchno. Issled. Inst. Tekhn. Mash., 33:26-37 (1962).

72. Böhme, Die zerstörungsfreie Prüfung von Aluminium-Werkstücken mit Ultraschallimpulsen [Nondestructive testing of aluminum articles with ultrasonic pulses], Aluminium (BRD), 34(4):200-205 (1958).

73. B. I. Vybornov, Application of surface and normal modes for the inspection of components, Materials of the Seminar "Physical Methods of Nondestructive Production Quality Control," No. 1, Moscow (1962).

74. B. I. Vybornov and K. A. Ogurtsov, The small size UZDL-61 ultrasonic defectoscope for testing of gas turbine blades, Zav. Lab., 28(8):997-998 (1962).

75. I. S. Ross, S. I. Kapuseienski, and K. B. Danields, Variable delay line using ultrasonic surface waves, Inst. Radio Engrs. Nat. Conv. Record, 6(2):118-120 (1958).

76. I. E. May, New trends in ultrasonic device technology, Report K-31, Fourth International Congress on Acoustics, Copenhagen, August 1962.

77. W. P. Mason, Uses of ultrasonics in radio, radar and sonar systems, Proc. Inst. Radio Engrs. 50(5):1374-1384, May (1962).

78. T. R. Meeker, Dispersive ultrasonic delay lines using the first longitudinal mode in a strip, Inst. Radio Engrs. Trans. Ultrasonics Eng., UE-7(2):53-58 (1960).

79. T. R. Meeker, The application of the theory of elastic waves in plates to the design of ultrasonic dispersive delay lines, Inst. Radio Engrs. Intern. Conv. Record, 9, Part 6:327-333 (1961).

80. A. H. Meitzler, Transmission characteristics of longitudinal-mode strip delay lines having asymmetrically tapered widths. Inst. Radio Engrs. Intern. Conv. Record, 9, Part 6:334-342 (1961).

81. A. H. Fitch, Synthesis of dispersive delay characteristics by thickness tapering in ultrasonic strip dealy lines, J. Acoust. Soc. Am., 33(11):1658 (1961).

82. I. R. Frederick, Use of ultrasonic surface waves in the determination of residual stress in metals, J. Acoust. Soc. Am., 32(11):1499 (1960).

83. P. Greguss, On the relation between the ultrasonic velocity and the surface properties of metals, Proceedings of Vibration Problems (Polish Academy of Sciences), 5:3-10 (1960).

84. F. J. Lockett, Effect of thermal properties of a solid on the velocity of Rayleigh waves, J. Mech. Phys. Solids, 7(1):71-75 (1958).

85. N. N. Egorov, Damping of Rayleigh waves in an elastic layer above a half-space, Akust. Zh., 7(3):378-380 (1961).

86. N. N. Egorov, Feasibility study of the application of ultrasonic surface waves for measuring the depth of certain types of surface-hardened layers, Collection: Application of Ultrasonics to the Investigation of Matter, Moscow (1961), No. 14, pp. 197-208.

87. C. L. Frederick and D. C. Worlton, Ultrasonic thickness measurements with Lamb waves, Nondstructive Testing, 20(1):51-55 (1962).

INDEX

GPSR Compliance

The European Union's (EU) General Product Safety Regulation (GPSR) is a set of rules that requires consumer products to be safe and our obligations to ensure this.

If you have any concerns about our products, you can contact us on ProductSafety@springernature.com

In case Publisher is established outside the EU, the EU authorized representative is:

Springer Nature Customer Service Center GmbH
Europaplatz 3
69115 Heidelberg, Germany

Batch number: 10854084

Printed by Printforce, the Netherlands